The Writer's Library

Science and Society

The HarperCollins Editors
with
Josephine Koster Tarvers

HarperCollins *CollegePublishers*

Acquisitions Editor: Patricia Rossi
Series Editor: David Munger
Cover Design: Stacy Agin
Production Administrator: Linda Murray
Printer and Binder: R. R. Donnelley & Sons Company
Cover Printer: The Lehigh Press, Inc.

For permission to use copyrighted material, grateful acknowledgement is made to the copyright holders on page 259 which is hereby made part of the copyright page.

Library of Congress Cataloging-in-Publication Data

Science and Society.
p. cm. -- (The Writer's Library)
ISBN 0-06-501126-0
1. Science--Philosophy. 2. Science--Moral and ethical aspects. 3. Science--Social aspects. 4. Technology--Philosophy.
Q175.S4174 1992
303.48'3--dc20 92-27835
CIP

Contents

1 What is Science?

2 Science vs. Pseudoscience

✧ ✧ ✧

3 The Connection of Humanity and Nature

✧ ✧ ✧

4 Science and Ethics

✧ ✧ ✧

5 The Costs and Benefits of Medical Technology

6 The Future of Science and Technology

Rhetorical Contents

✧

✧

✧

Narrative

✧ ✧ ✧

Description/Example

✧ ✧ ✧

Classification

✧ ✧ ✧

Comparison/Contrast

✧ ✧ ✧

Cause/Effect

✧ ✧ ✧

Definition

✧ ✧ ✧

Argument

✧ ✧ ✧

Researched Writing

✧ ✧ ✧

Preface

The Series

As editors, we see dozens of proposals for new readers every year. Each proposal is unique because it represents the thinking of an instructor at a different school, with different ideas about how to teach freshman composition. And every year, we publish several new readers based on those differing ideas, hoping composition programs will be able to adapt their own theories about teaching writing to the strategies advocated in our books. Unfortunately, if the teaching strategy of a given composition program does not closely match the available texts, the inevitable result is compromise.

The Writer's Library represents a revolutionary change from that type of thinking. Rather than choosing a text which forces them to adapt to its style of teaching, programs using *The Writer's Library* adapt *the text* to suit *their needs. The Writer's Library* is unique among readers for freshman composition because it is a flexible series of books, rather than a single, restrictive text. By using only the volumes they need, instructors can gain flexibility and save their students money—each volume of *The Writer's Library* is less than one-third the price of the average freshman composition reader.

Each volume of *The Writer's Library* covers a single theme, and is divided into chapters—sub-themes and issues that interest students. Each chapter has an introduction which contextualizes and demonstrates the relationships between readings. Each reading includes an introduction with a brief author biography, and suggestions for discussion and writing. An instructor's manual with suggested teaching strategies and suggested responses to discussion and writing questions is available.

The Writer's Library is flexible enough to be used in almost any composition classroom. Instructors who want to use a rhetoric or a handbook and supplement it with a few readings can choose one volume from the series at minimal additional cost to their students. Instructors who place a strong emphasis on reading can choose the two or three volumes that interest them—saving their students from purchasing lengthy readers and then using less than half of the selections.

This Volume

When we asked writing instructors what subjects we should cover in *The Writer's Library*, *Science and Society* received enthusiastic and widespread support. Instructors felt a text like this would help students realize how important science and technology had become in their lives, and would find discussing the questions raised by such a collection of essays stimulating and challenging.

This volume is organized into chapters that cover many facets of our subject: the definitions of science, "pseudo-scientific" theories, the subjective nature of interpretation, ethics, the health care system in America, and the future of science and technology. Instructors can follow the selections in the order they are presented, skip around in the text, or even alternate the readings in this volume with selections from other volumes in *The Writer's Library*.

Acknowledgements

The production of this series was a team effort, combining the varied talents of publishing professionals and writing instructors. We applied the ideas of dozens of writing instructors to shape the texts, and then worked with Judith Olson-Fallon and Josephine Koster Tarvers, experienced writing instructors both, to create the final product. Along the way, we received assistance from many talented individuals. Mark Gerrard contributed many suggestions while copy editing the final manuscript. Maria Paone was always there with a suggestion or wry remark. Tom Maeglin added input while obtaining permissions. But by far the most copious and helpful suggestions came from the instructors who helped us shape the original concept or reviewed draft manuscripts: Chris Anson, University of Minnesota; Thomas Blues, University of Kentucky; Mary Buckalew, North Texas State University; Marianne Cooley, University of Houston; George Gadda, UCLA; John Gage, University of Oregon; Anne Greene, Wesleyan University; Elizabeth Hodges, Virginia Commonwealth University; William Ingram, University of Michigan; Jim Killingsworth, Texas A&M University; Carl Klaus, University of Iowa; Barry Kroll, Indiana University; Bruce Leland, Western Illinois University; David Jolliffe, University of Illinois at Chicago; Russell Larson, Eastern Michigan University; Jay Ludwig, Michigan State University; Anne Matthews, Princeton University; George Miller, University of Delaware; Mark Patterson, University of Washington; Robert Perrin, Indiana State University; Paul Ranieri, Ball State University; Ruth Ray, Wayne State University; Tom Recchio, University of Connecticut; Kelly Reed, Northeastern University; Todd Sammons, University of Hawaii at Manoa; Charles I. Schuster, University of Wisconsin-Milwaukee; Joyce Smoot, Virginia Tech; Joyce Stauffer, Indiana University-Purdue; Irwin Weiser, Purdue University; Mark Wiley, California State University-Long Beach; and Richard Zbaracki, Iowa State University.

David Munger
Patricia Rossi
Laurie Likoff
Jane Kinney

To the Student

Science is a subject that polarizes its audiences: people either seem to love it or hate it. This may have something to do with the way you were taught about science, the people and attitudes with which you grew up, the value placed on science by your community, or a number of other factors. But whether the word "science" conjures up a picture of excitement in the laboratory or of Dr. Jekyll and Mr. Hyde, no one can deny that as we face the twenty-first century, all of us must come to terms with the roles science plays in our lives.

This book can help you do so by offering readings you can think, talk, and write about with your classmates. We begin with basic definitions: at the end of the twentieth century, what *is* science? And by contrast, what is pseudo-science, or not scientific at all? Next, we explore how science is connected to the natural world around us, and how we as a society have forged that connection. Questions of science and society lead us to ethics, the boundaries we place on scientific activity and that science places on itself. Finally, we conclude with two sections that examine the practical aspects of science and society: the costs and benefits of scientific technology, and the ways science may be shaping and changing our futures.

Each essay is accompanied by a brief note providing information about the author and the original context for the work. And each is followed by a group of questions about the work, asking you to think about not only the reading's content but its rhetorical techniques. Some of these you may answer alone, some you may work out with classmates, and some you may be asked to write about for a grade. Some will ask you to conduct research to gain a more complete picture. Of course, there are many more questions to be asked about each essay than we can accommodate in this framework; we encourage you to explore these other avenues on your own.

The theologian Pierre Teilhard de Chardin, writing of the vast advances in science and technology in our lifetimes, asked, "The future of the earth is in our hands: How shall we decide?" Knowledge, he concluded, is the key. By studying the relations of science and society, we hope you will be able to shape a safer, more profitable future for yourself and the members of your community.

1 What is Science?

What is science? Chances are, your answer will say that it is something about facts, and definite answers, and men in white coats. It's likely that you will try to cast your answer in terms of things: experiments, chemicals, equations, equipment. There may even be mentions of Dr. Frankenstein or television programs or nerds. Science strikes most readers as a "cut-and-dried" state, not an evolving pursuit.

The essays in this section, by some of our century's finest scientists and writers, seek to dispel those stereotypes. They speak of science as a search, as a continual questioning, as a journey without end. All would agree with Alfred Church Lane's famous dictum that "The larger the area of our (scientific) knowledge, the greater the circumstance of our ignorance." Here, then, are five attempts to lessen that ignorance. Jacob Bronowski describes science as an act of artistic creation; Lewis Thomas casts it as a continual challenge to our assumptions of certainty. Horace Freeland Judson knowingly describes the scientist's obsession with finding new answers, and J. Robert Oppenheimer sees it as the ultimate expression of love. Finally, Albert Einstein, perhaps the greatest scientist in many centuries, shows how science, in even its simplest equations, can be an instrument that leads to beautiful and moral consequences. As you read these essays, ask yourself, "How would I define science? How has my definition been formed? What do I know about it? What have I been missing?"

Jacob Bronowski

The Nature of Scientific Reasoning

The distinguished mathematician and author JACOB BRONOWSKI (pronounced *Bron-OFF-ski*) was born in Poland in 1908. His academic career took him to positions in England, a UN job studying the effects of the atomic bombing of Nagasaki, and to many prestigious academic appointments in the United States before his death in 1974. But to most Americans he is best known as a television personality. His 1973 BBC series *The Ascent of Man* (and the book of the same title) was shown on public television in the US to great popular acclaim; it celebrated the development of humanity's attempts to understand and control nature, from prehistory to the present. John Lenihan wrote of the series, "Bronowski saw science as a part of man's cultural heritage; to him the progress of science was. . . a shifting pattern which could be appreciated only by recognising the interwoven strands of history, art, literature, and philosophy."

Bronowski summed up his chief themes in an interview: "Science and art are wonderfully human because they both call on imagination and they both require enormous dedication and integrity. . . . If you care about art or if you care about science you must have a huge sense of involvement with what is human about those things." This essay first appeared in his book *Science and Human Values* in 1956.

What is the insight in which the scientist tries to see into nature? Can it indeed be called either imaginative or creative? To the literary man the question may seem merely silly. He has been taught that science is a large collection of facts; and if this is true, then the only seeing which scientists need to do is, he supposes, seeing the facts. He pictures them, the colorless professionals of science, going off to work in the morning into the universe in a neutral, unexposed state. They then expose themselves like a photographic plate. And then in the darkroom or laboratory they develop the image, so that suddenly and startlingly it appears, printed in capital letters, as a new formula for atomic energy.

2 Men who have read Balzac and Zola are not deceived by the claims of these writers that they do no more than record the facts. The readers of Christopher Isherwood do not take him literally when he writes "I am a camera." Yet the same readers solemnly carry with them from their schooldays this foolish picture of the scientist fixing by some mechanical process the facts of nature. I have had of all people a historian tell me that science is a collection of facts, and his voice had not even the ironic rasp of one filing cabinet reproving another.

3 It seems impossible that this historian had ever studied the beginnings of a scientific discovery. The Scientific Revolution can be held to begin in the year 1543 when there was brought to Copernicus, perhaps on his deathbed, the first printed copy of the book he had finished about a dozen years earlier. The thesis of this book is that the earth moves around the sun. When did Copernicus go out and record this fact with his camera? What appearance in nature prompted his outrageous guess? And in what odd sense is this guess to be called a neutral record of fact?

4 Less than a hundred years after Copernicus, Kepler published (between 1609 and 1619) the three laws which describe the paths of the planets. The work of Newton and with it most of our mechanics spring from these laws. They have a solid, matter-of-fact sound. For example, Kepler says that if one squares the year of a planet, one gets a number which is proportional to the cube of its average distance from the sun. Does anyone think that such a law is found by taking enough readings and then squaring and cubing everything in sight? If he does, then, as a scientist, he is doomed to a wasted life; he has as little prospect of making a scientific discovery as an electronic brain has.

5 It was not this way that Copernicus and Kepler thought, or that scientists think today. Copernicus found that the orbits of the planets would look simpler if they were looked at from the sun and not from the earth. But he did not in the first place find this by routine calculation. His first step was a leap of imagination—to lift himself from

the earth, and put himself wildly, speculatively into the sun. "The earth conceives from the sun," he wrote; and "the sun rules the family of stars." We catch in his mind an image, the gesture of the virile man standing in the sun, with arms outstretched, overlooking the planets. Perhaps Copernicus took the picture from the drawings of the youth with outstretched arms which the Renaissance teachers put into their books on the proportions of the body. Perhaps he had seen Leonardo's drawings of his loved pupil Salai. I do not know. To me, the gesture of Copernicus, the shining youth looking outward from the sun, is still vivid in a drawing which William Blake in 1780 based on all these: the drawing which is usually called *Glad Day*.

6 Kepler's mind, we know, was filled with just such fanciful analogies; and we know what they were. Kepler wanted to relate the speeds of the planets to the musical intervals. He tried to fit the five regular solids into their orbits. None of these likenesses worked, and they have been forgotten, yet they have been and they remain the stepping stones of every creative mind. Kepler felt for his laws by way of metaphors, he searched mystically for likenesses with what he knew in every strange corner of nature. And when among these guesses he hit upon his laws, he did not think of their numbers as the balancing of a cosmic bank account, but as a revelation of the unity in all nature. To us, the analogies by which Kepler listened for the movement of the planets in the music of the spheres are farfetched. Yet are they more so than the wild leap by which Rutherford and Bohr in our own century found a model for the atom in, of all places, the planetary system?

7 No scientific theory is a collection of facts. It will not even do to call a theory true or false in the simple sense in which every fact is either so or not so. The Epicureans held that matter is made of atoms two thousand years ago and we are now tempted to say that their theory was true But if we do so we confuse their notion of matter with our own. John Dalton in 1808 first saw the structure of matter as we do today, and what he took from the ancients was not their theory but something richer, their image: the atom. Much of what was in Dalton's mind was as vague as the Greek notion, and quite as mistaken. But he suddenly gave life to the new facts of chemistry and the ancient theory together, by fusing them to give what neither had: a coherent picture of how matter is linked and built up from different kinds of atoms. The act of fusion is the creative act.

8 All science is the search for unity in hidden likenesses. The search may be on a grand scale, as in the modern theories which try to link the fields of gravitation and electromagnetism. But we do not need to be browbeaten by the scale of science. There are discoveries to be made by snatching a small likeness from the air too, if it is bold enough. In 1935 the Japanese physicist Hideki Yukawa wrote a paper which can still give heart to a young scientist. He took as his starting point the known fact that waves of light can sometimes behave as if they were separate pellets. From this he reasoned that the forces which hold the nucleus of an atom together might sometimes also be observed as if they were solid pellets. A schoolboy can see how thin Yukawa's analogy is, and his teacher would be severe with it. Yet Yukawa without a blush calculated the mass of the pellet he expected to see, and waited. He was right; his meson was found, and a range of other mesons, neither the existence nor the nature of which had been suspected before. The likeness had borne fruit.

9 The scientist looks for order in the appearances of nature by exploring such likenesses. For order does not display itself of itself; if it can be said to be there at all, it is not there for the mere looking. There is no way of pointing a finger or camera at it; order must be discovered and, in a deep sense, it must be created. What we see, as we see it, is mere disorder.

10 This point has been put trenchantly in a fable by Karl Popper. Suppose that someone wished to give his whole life to science. Suppose that he therefore sat down, pencil in hand, and for the next twenty, thirty, forty years recorded in notebook after notebook everything that he could observe. He may be supposed to leave out nothing: today's humidity, the racing results, the level of cosmic radiation and the stockmarket prices and the look of Mars, all would be there. He would have compiled the most careful record of nature that has ever been made; and, dying in the calm certainty of a life well spent, he would of course leave his notebooks to the Royal Society. Would the Royal Society thank him for the treasure of a lifetime of observation? It would not. The Royal Society would treat his notebooks exactly as the English bishops have treated Joanna Southcott's box. It would refuse to open them at all, because it would know without looking that the notebooks contain only a jumble of disorderly and meaningless items.

11 Science finds order and meaning in our experience, and sets about this in quite a different way. It sets about it as Newton did in the story which he himself told in his old age, and of which the schoolbooks give only a caricature. In the year 1665, when Newton was twenty-two, the plague broke out in southern England, and the University of Cambridge was closed. Newton therefore spent the next eighteen months at home, removed from traditional learning, at a time when he was impatient for knowledge and, in his own phrase, "I was in the prime of my age for invention." In this eager, boyish mood, sitting one day in the garden of his widowed mother, he saw an apple fall. So far the books have the story right; we think we even know the kind of apple; tradition has it that it was a Flower of Kent. But now they miss the crux of the story. For what struck the young Newton at the sight was not the thought that the apple must be drawn to the earth by gravity; that conception was older than Newton. What struck him was the conjecture that the same force of gravity which reaches to the top of the tree, might go on reaching out beyond the earth and its air, endlessly into space. Gravity might reach the moon: this was Newton's new thought; and it might be gravity which holds the moon in her orbit. There and then he calculated what force from the earth (falling off as the square of the distance) would hold the moon, and compared it with the known force of gravity at tree height. The forces agreed; Newton says laconically, "I found the answer pretty nearly." Yet they agreed only nearly: the likeness and the approximation go together, for no likeness is exact. In Newton's science modern science is full grown.

12 It grows from a comparison. It has seized a likeness between two unlike appearances; for the apple in the summer garden and the grave moon overhead are surely as unlike in their movements as two things can be. Newton traced in them two expressions of a single concept, gravitation: and the concept (and the unity) are in that sense his free creation. The progress of science is the discovery at each step of a new order which gives unity to what had long seemed unlike.

Suggestions for Discussion and Writing

1. Bronowski begins by making an analogy between scientists and unexposed photographic film. Do you think this analogy works? What other analogies can you draw that might

describe your preconceptions and knowledge about scientists?

2. For most of us, Bronowski's assertion that science does not consist of taking readings and measurements is shocking. What instead does Bronowski think science consists of? In your opinion, why do scientists spend time in laboratories if measurements are unimportant?

3. One principle of science most people remember from high school is *entropy*, the principle that all systems in nature tend to disorder. If this principle is true, is it possible for "scientists (to) find order and meaning in our experience"? How?

4. Bronowski uses many comparisons in his essay. Identify as many of them as you can. What effect does he hope to have on his readers by using this technique?

5. Interview scientists and science majors you know or can find on your campus. How would they define what science is? What a scientist does? How do their definitions fit with Bronowski's claims?

Lewis Thomas

Debating the Unknowable

The essayist, critic, physician and scientist LEWIS THOMAS (born 1913) has received numerous honors and awards for his multiple careers. Born in Flushing, New York, he trained in New England and eventually became head of Memorial-Sloan Kettering Cancer Research Center in New York City. He gained recognition as a writer with *Lives of a Cell* (1974), for which he won the National Book Award; this has been followed by *The Medusa and the Snail* (1979), *The Youngest Science* (1983), *Late Night Thoughts on Listening to Mahler's Ninth Symphony* (1983), and *Et cetera, Et cetera: Notes of a Word-Watcher* (1990).

A review of *Late Night Thoughts* summed up

Thomas' work thus: "Thomas is deeply worried about the future of mankind. He can accept death as a part of the natural order, but not the death of everything. Like many of us, he is baffled by how we could have got ourselves to the present impasse. Unlike most of us, he can put his concerns into language that gives them a special depth and urgency." This is a selection from *Late Night Thoughts.*

The greatest of all the accomplishments of twentieth-century science has been the discovery of human ignorance. We live, as never before, in puzzlement about nature, the universe, and ourselves most of all. It is a new experience for the species. A century ago, after the turbulence caused by Darwin and Wallace had subsided and the central idea of natural selection had been grasped and accepted, we thought we knew everything essential about evolution. In the eighteenth century there were no huge puzzles; human reason was all you needed in order to figure out the universe. And for most of the earlier centuries, the Church provided both the questions and the answers, neatly packaged. Now, for the first time in human history, we are catching glimpses of our incomprehension. We can still make up stories to explain the world, as we always have, but now the stories have to be confirmed and reconfirmed by experiment. This is the scientific method, and once started on this line we cannot turn back. We are obliged to grow up in skepticism, requiring proofs for every assertion about nature, and there is no way except to move ahead and plug away, hoping for comprehension in the future but living in a condition of intellectual instability for the long time.

2 It is the admission of ignorance that leads to progress, not so much because the solving of a particular puzzle leads directly to a new piece of understanding but because the puzzle—if it interests enough scientists—leads to *work.* There is a similar phenomenon in entomology known as stigmergy, a term invented by Grassé, which means "to incite to work." When three or four termites are collected together in a chamber they wander about aimlessly, but when more termites are added, they begin to build. It is the presence of other termites, in sufficient numbers at close quarters, that produces the work: they pick up each other's fecal pellets and stack them in neat columns, and when the columns are precisely the right height, the termites reach across and

turn the perfect arches that form the foundation of the termitarium. No single termite knows how to do any of this, but as soon as there are enough termites gathered together they become flawless architects, sensing their distances from each other although blind, building an immensely complicated structure with its own air-conditioning and humidity control. They work their lives away in this ecosystem built by themselves. The nearest thing to a termitarium that I can think of in human behavior is the making of language, which we do by keeping *at* each other all our lives, generation after generation, changing the structure by some sort of instinct.

3 Very little is understood about this kind of collective behavior. It is out of fashion these days to talk of "superorganisms," but there simply aren't enough reductionist details in hand to explain away the phenomenon of termites and other social insects: some very good guesses can be made about their chemical signaling systems, but the plain fact that they exhibit something like a collective intelligence is a mystery, or anyway an unsolved problem, that might contain important implications for social life in general. This mystery is the best introduction I can think of to biological science in college. It should be taught for its strangeness, and for the ambiguity of its meaning. It should be taught to premedical students, who need lessons early in their careers about the uncertainties in science.

4 College students, and for that matter high school students, should be exposed very early, perhaps at the outset, to the big arguments currently going on among scientists. Big arguments stimulate their interest, and with luck engage their absorbed attention. Few things in life are as engrossing as a good fight between highly trained and skilled adversaries. But the young students are told very little about the major disagreements of the day; they may be taught something about the arguments between Darwinians and their opponents a century ago, but they do not realize that similar disputes about other matters, many of them touching profound issues for our understanding of nature, are still going on and, indeed, are an essential feature of the scientific process. There is, I fear, a reluctance on the part of science teachers to talk about such things, based on the belief that before students can appreciate what the arguments are about they must learn and master the "fundamentals." I would be willing to see some experiments along this line, and I have in mind several examples of contemporary doctrinal dispute in which the drift of the argument can be readily per-

ceived without deep or elaborate knowledge of the subject.

5 There is, for one, the problem of animal awareness. One school of ethologists devoted to the study of animal behavior has it that human beings are unique in the possession of consciousness, differing from all other creatures in being able to think things over, capitalize on past experience, and hazard informed guesses at the future. Other, "lower," animals (with possible exceptions made for chimpanzees, whales, and dolphins) cannot do such things with their minds; they live from moment to moment with brains that are programmed to respond, automatically or by conditioning, to contingencies in the environment. Behavioral psychologists believe that this automatic or conditioned response accounts for human mental activity as well, although they dislike that word "mental." On the other side are some ethologists who seem to be more generous-minded, who see no compelling reasons to doubt that animals in general are quite capable of real thinking and do quite a lot of it—thinking that isn't as dense as human thinking, that is sparser because of the lack of language and the resultant lack of metaphors to help the thought along, but thinking nonetheless.

6 The point about this argument is not that one side or the other is in possession of a more powerful array of convincing facts; quite the opposite. There are not enough facts to sustain a genuine debate of any length; the question of animal awareness is an unsettled one. In the circumstance, I put forward the following notion about a small beetle, the mimosa girdler, which undertakes three pieces of linked, sequential behavior: finding a mimosa tree and climbing up the trunk and out to the end of a branch; cutting a longitudinal slit and laying within it five or six eggs; and crawling back on the limb and girdling it neatly down into the cambium. The third step is an eight-to-ten-hour task of hard labor, from which the beetle gains no food for itself—only the certainty that the branch will promptly die and fall to the ground in the next brisk wind, thus enabling the larvae to hatch and grow in an abundance of dead wood. I propose, in total confidence that even though I am probably wrong nobody today can prove that I am wrong, that the beetle is not doing these three things out of blind instinct, like a little machine, but is thinking its way along, just as we would think. The difference is that we possess enormous brains, crowded all the time with an infinite number of long thoughts, while the beetle's brain is only a few strings of neurons connected in a

modest network, capable therefore of only three tiny thoughts, coming into consciousness one after the other: find the right tree; get up there and lay eggs in a slit; back up and spend the day killing the branch so the eggs can hatch. End of message. I would not go so far as to anthropomorphize the mimosa tree, for I really do not believe plants have minds, but something has to be said about the tree's role in this arrangement as a beneficiary: mimosas grow for twenty-five to thirty years and then die, unless they are vigorously pruned annually, in which case they can live to be a hundred. The beetle is a piece of good luck for the tree, but nothing more: one example of pure chance working at its best in nature—what you might even wish to call good nature.

7 This brings me to the second example of unsettlement in biology, currently being rather delicately discussed but not yet argued over, for there is still only one orthodoxy and almost no opposition, yet. This is the matter of chance itself and the role played by blind chance in the arrangement of living things on the planet. It is, in the orthodox view, pure luck that evolution brought us to our present condition, and things might just as well have turned out any number of other, different ways, and might go in any unpredictable way for the future. There is, of course, nothing chancy about natural selection itself: it is an accepted fact that selection will always favor the advantaged individuals whose genes succeed best in propagating themselves within a changing environment. But the creatures acted upon by natural selection are themselves there as the result of chance: mutations (probably of much more importance during the long period of exclusively microbial life starting nearly 4 billion years ago and continuing until about one billion years ago); the endless sorting and resorting of genes within chromosomes during replication; perhaps recombination of genes across species lines at one time or another; and almost certainly the carrying of genes by viruses from one creature to another.

8 The argument comes when one contemplates the whole biosphere, the conjoined life of the earth. How could it have turned out to possess such stability and coherence, resembling as it does a sort of enormous developing embryo with nothing but chance events to determine its emergence? Lovelock and Margulis, facing this problem, have proposed the Gaia Hypothesis, which is, in brief, that the earth is itself a form of life, "a complex entity involving the Earth's bio-

sphere, atmosphere, oceans and soil; the totality constituting a feedback or cybernetic system which seeks an optimal physical and chemical environment for life on this planet." Lovelock postulates, in addition, that "the physical and chemical condition of the surface of the Earth, of the atmosphere, and of the oceans has been and is actively made fit and comfortable by the presence of life itself."

9 This notion is beginning to stir up a few signs of storm, and if it catches on, as I think it will, we will soon find the biological community split into fuming factions, one side saying that the evolved biosphere displays evidences of design and purpose, the other decrying such heresy. I believe that students should learn as much as they can about the argument. In an essay in *Coevolution* (Spring 1981), W. F. Doolittle has recently attacked the Gaia Hypothesis, asking, among other things, ". . .how does Gaia know if she is too cold or too hot and how does she instruct the biosphere to behave accordingly?" This is not a deadly criticism in a world where we do not actually understand, in anything like real detail, how even Dr. Doolittle manages the stability and control of his own internal environment, including his body temperature. One thing is certain: none of us can instruct our body's systems to make the needed corrections beyond a very limited number of rather trivial tricks made possible through biofeedback techniques. If something goes wrong with my liver or my kidneys, I have no advice to offer out of my cortex. I rely on the system to fix itself, which it usually does with no help from me beyond crossing my fingers.

10 Another current battle involving the unknown is between sociobiologists and antisociobiologists, and it is a marvel for students to behold. To observe, in open-mouthed astonishment, one group of highly intelligent, beautifully trained, knowledgeable, and imaginative scientists maintaining that all behavior, animal and human, is governed exclusively by genes, and another group of equally talented scientists asserting that all behavior is set and determined by the environment or by culture, is an educational experience that no college student should be allowed to miss. The essential lesson to be learned has nothing to do with the relative validity of the facts underlying the argument. It is the argument itself that is the education: we do not yet know enough to settle such questions.

11 One last example. There is an uncomfortable secret in biology, not much talked about yet, but beginning to surface. It is, in a way,

linked to the observations that underlie the Gaia Hypothesis. Nature abounds in instances of cooperation and collaboration, partnerships between species. There is a tendency of living things to join up whenever joining is possible: accommodation and compromise are more common results of close contact than combat and destruction. Given the opportunity and the proper circumstances, two cells from totally different species—a mouse cell and a human cell, for example—will fuse to become a single cell, and then the two nuclei will fuse into a single nucleus, and then the hybrid cell will divide to produce generations of new cells containing the combined genomes of both species. Bacteria are indispensable partners in the fixation of atmospheric nitrogen by plants. The oxygen in our atmosphere is put there, almost in its entirety, by the photosynthetic chloroplasts in the cells of green plants, and these organelles are almost certainly the descendants of blue-green algae that joined up when the nucleated cells of higher plants came into existence. The mitochondria in all our own cells, and in all other nucleated cells, which enable us to use oxygen for energy, are the direct descendants of symbiotic bacteria. These are becoming accepted facts, and there is no longer an agitated argument over their probable validity; but there are no satisfactory explanations for how such amiable and useful arrangements came into being in the first place. Axelrod and Hamilton (*Science,* March 27, 1981) have recently reopened the question of cooperation in evolution with a mathematical approach based on game theory (the Prisoner's Dilemma game), which permits the hypothesis that one creature's best strategy for dealing repeatedly with another is to concede and cooperate rather than to defect and go it alone.

12 This idea can be made to fit with the mathematical justification based on kinship already accepted for explaining altruism in nature—that in a colony of social insects the sacrifice of one individual for another depends on how many of the sacrificed member's genes are matched by others and thus preserved, and that the extent of the colony's altruistic behavior can be mathematically calculated. It is, by the way, an interesting aspect of contemporary biology that true altruism—the giving away of something without return—is incompatible with dogma, even though it goes on all over the place. Nature, in this respect, keeps breaking the rules, and needs correcting by new ways of doing arithmetic.

13 The social scientists are in the hardest business of all—trying to

understand how humanity works. They are caught up in debates all over town; everything they touch turns out to be one of society's nerve endings, eliciting outrage and cries of pain. Wait until they begin coming close to the bone. They surely will someday, provided they can continue to attract enough bright people—fascinated by humanity, unafraid of big numbers, and skeptical of questionnaires—and provided the government does not starve them out of business, as is now being tried in Washington. Politicians do not like pain, not even wincing, and they have some fear of what the social scientists may be thinking about thinking for the future.

14 The social scientists are themselves too modest about the history of their endeavor, tending to display only the matters under scrutiny today in economics, sociology, and psychology, for example—never boasting, as they might about one of the greatest of all scientific advances in our comprehension of humanity, for which they could be claiming credit. I refer to the marvelous accomplishments of the nineteenth-century comparative linguists. When the scientific method is working at its best, it succeeds in revealing the connection between things in nature that seem at first totally unrelated to each other. Long before the time when the biologists, led by Darwin and Wallace, were constructing the tree of evolution and the origin of species, the linguists were hard at work on the evolution of language. After beginning in 1786 with Sir William Jones and his inspired hunch that the remarkable similarities among Sanskrit, Greek, and Latin meant, in his words, that these three languages must "have sprung from some common source, which, perhaps, no longer exists," the new science of comparative grammar took off in 1816 with Franz Bopp's classic work "On the conjugational system of the Sanskrit language in comparison with that of the Greek, Latin, Persian and Germanic languages"—a piece of work equivalent, in its scope and in its power to explain, to the best of nineteenth-century biology. The common Indo-European ancestry of English, Germanic, Slavic Greek, Latin, Baltic, Indic, Iranian, Hittite, and Anatolian tongues, and the meticulous scholarship connecting them was a tour de force for research—science at its best, and social science at that.

15 It is nice to know that a common language, perhaps 20,000 years ago, had a root word for the earth which turned, much later, into the technical term for the complex polymers that make up the connective tissues of the soil: humus and what are called the humic acids. There is

a strangeness, though, in the emergence from the same root of words such as "human" and "humane," and "humble." It comes as something of a shock to realize that the root for words such as "miracle" and "marvel" meant, originally, "to smile," and that from the single root *sa* were constructed, in the descendant tongues, three cognate words, "satisfied," "satiated," and "sadness." How is it possible for a species to show so much wisdom in its most collective of all behaviors—the making and constant changing of language—and at the same time be so habitually folly-prone in the building of nation-states? Modern linguistics has moved into new areas of inquiry as specialized and inaccessible for most laymen (including me) as particle physics; I cannot guess where linguistics will come out, but it is surely aimed at scientific comprehension, and its problem—human language—is as crucial to the species as any other field I can think of, including molecular genetics.

16 But there are some risks involved in trying to do science in the humanities, before its time, and useful lessons can be learned from some of the not-so-distant history of medicine. A century ago it was the common practice to deal with disease by analyzing what seemed to be the underlying mechanism and applying whatever treatment popped into the doctor's head. Getting sick was a hazardous enterprise in those days. The driving force in medicine was the need to *do* something, never mind what. It occurs to me now, reading in incomprehension some of the current reductionist writings in literary criticism, especially poetry criticism, that the new schools are at risk under a similar pressure. A poem is a healthy organism, really in need of no help from science, no treatment except fresh air and exercise. I thought I'd just sneak that in.

Suggestions for Discussion and Writing

1. Where Bronowski argued that science stresses certainty, Thomas argues that science stresses *un*certainty. Can their two perspectives be reconciled? How would you do it?

2. According to Thomas, what's wrong with how we teach science? What methods would he use to replace current ones?

3. Explain the organizational structure(s) that Thomas uses to convey his message. Why do you think he chose such an organization?

4. What effect does the conclusion have on you as readers? Why do you think Thomas chose to end the essay in this manner?

5. What science teaching methods are used on your campus? Do any of your teachers use "problem-solving" or "alternate-path" teaching methods? Interview some of your teachers to find out how and why they chose the teaching methods they use. What conclusions can you draw?

Horace Freeland Judson
The Rage to Know

Born in New York City in 1931, HORACE FREELAND JUDSON has made a career of writing the histories of science and its leading figures. He describes his methods thus: "I pursue. . . materials, both oral and written, more intensively than most. I work for the most part with a tape recorder. . . and transcribe the interviews myself. (Then I give the transcript to the subject and. . .) I interview him again, going over the text page by page, if necessary, with the tape recorder running. (Sometimes the subject says,) 'It wasn't *quite* like that—.' And then you get the gold."

His books include *Heroin Addiction in Britain* (1974), *The Eighth Day of Creation: Makers of the Revolution in Biology* (1979), and an award-winning collection of interviews with modern scientists, *The Search for Solutions* (1980), from which this essay is excerpted.

Certain moments of the mind have a special quality of well-being. A mathematician friend of mine remarked the other day that his daughter, aged eight, had just stumbled without his teaching onto the fact that some numbers are prime numbers—those, like 11 or 19 or 83 or 1023, that cannot be divided by any other integer (except, trivially, by 1). "She called them 'unfair' numbers," he said. "And when I asked her why they were unfair, she told me, 'Because there's no way to share them out evenly.'" What delighted him most was not her charming turn of phrase nor her equitable turn of mind (seventeen peppermints to give to her friends?) but—as a mathematician—the knowledge that the child had experienced a moment of pure scientific perception. She had discovered for herself something of the way things are.

2 The satisfaction of such a moment at its most intense—and this is what ought to be meant, after all, by the tarnished phrase "the moment of truth"—is not easy to describe. It partakes at once of exhilaration and tranquillity. It is luminously clear. It is beautiful. The clarity of the moment of discovery, the beauty of what in that moment is seen to be true about the world, is the fundamental attraction that draws scientists on.

3 Science is enormously disparate—easily the most varied and diverse of human pursuits. The scientific endeavor ranges from the study of animal behavior all the way to particle physics, and from the purest of mathematics back again to the most practical problems of shelter and hunger, sickness and war. Nobody has succeeded in catching all this in one net. And yet the conviction persists—scientists themselves believe, at heart—that behind the diversity lies a unity. In those luminous moments of discovery, in the various approaches and the painful tension required to arrive at them, and then in the community of science organized worldwide to doubt and criticize, test and exploit discoveries—somewhere in that constellation, to begin with, there are surely constants. Deeper is the lure that in the bewildering variety of the world as it is there may be found some astonishing simplicities.

4 Philosophers, and some of the greatest among them, have offered descriptions of what they claim is the method of science. These make most scientists acutely uncomfortable. The descriptions don't seem to fit what goes on in the doing of science. They seem at once too abstract and too limited. Scientists don't believe that they think in ways that are wildly different from the way most people think at least in some areas of their lives. "We'd be in real trouble—we could get nowhere—if ordinary methods of inference did not apply," Philip Morrison said in a conversation a while ago. (Morrison is a theoretical physicist at the Massachusetts Institute of Technology.) The wild difference, he went on to say, is that scientists apply these everyday methods to areas that most people never think about seriously and carefully. The philosophers' descriptions don't prepare one for either this

ordinariness or this extreme diversity of the scientific enterprise—the variety of things to think about, the variety of obstacles and traps to understanding, the variety of approaches to solutions. They hardly acknowledge the fact that a scientist ought often to find himself stretching to the tiptoe of available technique and apparatus, out beyond the frontier of the art, attempting to do something whose difficulty is measured most significantly by the fact that it has never been done before. Science is carried on—this, too, is obvious—in the field, in the observatory, in the laboratory. But historians leave out the arts of the chef and the watchmaker, the development at the bench of a new procedure or a new instrument. "And *making it work*," Morrison said. "This is terribly important." Indeed, biochemists talk about "the cookbook." Many a Nobel Prize has been awarded, not for a discovery, as such, but for a new technique or a new tool that opened up a whole field of discovery. "I am a theoretician," Morrison said. "And yet the most important problem for me is to be in touch with the people who are making new instruments or finding new ways of observing, and to try to get them to do the right experiments." And then, in a burst of annoyance, "I feel very reluctant to give any support to descriptions of 'scientific method.' The scientific enterprise is very difficult to model. You have to look at what scientists of all kinds *actually do*."

5

It's true that by contrast philosophers and historians seem book-bound—or paper-blindered, depending chiefly on what has been published as scientific research for their understanding of the process of discovery. In this century, anyway, published papers are no guide to the way scientists get the results they report. We have testimony of the highest authenticity for that. Sir Peter Medawar has both done fine science and written well about how it is done: he won his Nobel Prize for investigations of immunological tolerance, which explained, among other things, why foreign tissue, like a kidney or a heart, is rejected by the body into which it is transplanted, and he has described the methods of science in essays of grace and distinction. A while ago, Medawar wrote, "What scientists do has never been the subject of a scientific. . . inquiry. It is no use looking to scientific 'papers,' for they not merely conceal but actively misrepresent the reasoning that goes into the work they describe." The observation has become famous, its truth acknowledged by other scientists. Medawar wrote further, "Scientists are building explanatory structures, *telling stories* which are scrupulously tested to see if they are stories about real life."

6 Scientists do science for a variety of reasons, of course, and most of them are familiar to the sculptor, or to the surgeon or the athlete or the builder of bridges: the professional's pride in skill; the swelling gratification that comes with recognition accorded by colleagues and peers; perhaps the competitor's fierce appetite; perhaps ambition for a kind of fame more durable than most. At the beginning is curiosity, and with curiosity the delight in mastery—the joy of figuring it out that is the birthright of every child. I once asked Murray Gell-Mann, a theoretical physicist, how he got started in science. His answer was to point to the summer sky: "When I was a boy, I used to ask all sorts of simple questions—like, 'What holds the clouds up?'" Rosalind Franklin, the crystallographer whose early death deprived her of a share in the Nobel Prize for the discovery of the structure of DNA, one day was helping a young collaborator draft an application for research money, when she looked up at him and said, "What we can't tell them is that it's so much *fun!*" He still remembers her glint of mischief. The play of the mind in an almost childlike innocence, is a pleasure that appears again and again in scientists' reflections on their work. The geneticist Barbara McClintock, as a woman in American science in the 1930s, had no chance at the academic posts open to her male colleagues, but that hardly mattered to her. "I did it because it was *fun!*" she said forty years later. "I couldn't wait to get up in the morning! I never thought of it as 'science.' "

7 The exuberant innocence can be poignant. François Jacob, who won his share of a Nobel Prize as one of the small group of molecular biologists in the fifties who brought sense and order into the interactions by which bacteria regulate their life processes, recently read an account I had written of that work, and said to me with surprise and an evident pang of regret, "We were like children playing!" He meant the fun of it—but also the simplicity of the problems they had encountered and the innocence of mind they had brought to them. Two hundred and fifty years before—although Jacob did not consciously intend the parallel—Isaac Newton, shortly before his death, said:

> I do not know what I may appear to the world, but to myself I seem to have been only like a boy playing on the sea shore, and diverting myself in now and then finding a smoother pebble or a prettier shell than ordinary, whilst the great ocean of truth lay all undiscovered before me.

8 For some, curiosity and the delight of putting the world together deepen into a life's passion. Sheldon Glashow, a fundamental-particle physicist at Harvard, also got started in science by asking simple questions. "In eighth grade, we were learning about how the earth goes around the sun, and the moon around the earth, and so on," he said. "And I thought about that, and realized that the Man in the Moon is always looking at us"—that the moon as it circles always turns the same face to the earth. "And I asked the teacher, 'Why is the Man in the Moon always looking at us?' She was pleased with the question—but said it was hard to answer. And it turns out that it's not until you're in college-level physics courses that one really learns the answers," Glashow said. "But the *difference* is that most people would look at the moon and wonder for a moment and say, 'That's interesting'—and then forget it. But some people can't let go."

9 Curiosity is not enough. The word is too mild by far, a word for infants. Passion is indispensable for creation, no less in the sciences than in the arts. Medawar once described it in a talk addressed to young scientists. "You must feel in yourself an exploratory impulsion—an *acute discomfort* at incomprehension." This is the rage to know. The other side of the fun of science, as of art, is pain. A problem worth solving will surely require weeks and months of lack of progress, whipsawed between hope and the blackest sense of despair. The marathon runner or the young swimmer who would be a champion knows at least that the pain may be a symptom of progress. But here the artist and the scientist part company with the athlete—to join the mystic for a while. The pain of creation, though not of the body, is in one way worse. It must be not only endured but reflected back on itself to increase the agility, variety, inventiveness of the play of the mind. Some problems in science have demanded such devotion, such willingness to bear repeated rebuffs, not just for years but for decades. There are times in the practice of the arts, we're told, of abysmal self-doubt. There are like passages in the doing of science.

10 Albert Einstein took eleven years of unremitting concentration to produce the general theory of relativity; long afterward, he wrote, "In the light of knowledge attained, the happy achievement seems almost a matter of course, and any intelligent student can grasp it without too much trouble. But the years of anxious searching in the dark, with their intense longing, their alternations of confidence and exhaustion, and the final emergence into the light—only those who have

experienced it can understand it." Einstein confronting Einstein's problems: the achievement, to be sure, is matched only by Newton's and perhaps Darwin's—but the experience is not rare. It is all but inseparable from high accomplishment. In the black cave of unknowing, when one is groping for the contours of the rock and the slope of the floor, tossing a pebble and listening for its fall, brushing away false clues as insistent as cobwebs, a touch of fresh air on the cheek can make hope leap up, an unexpected scurrying whisper can induce the mood of the brink of terror. "Afterward it can be told—trivialized—like a *roman policier*, a detective story," François Jacob once said. "While you're there, it is the sound and the fury." But it was the poet and adept of mysticism St. John of the Cross who gave to this passionate wrestling with bafflement the name by which, ever since, it has been known: "the dark night of the soul."

11 Enlightenment may not appear, or not in time; the mystic at least need not fear forestalling. Enlightenment may dawn in ways as varied as the individual approaches of scientists at work—and, in defiance of stereotypes, the sciences far outrun the arts in variety of personal styles and in the crucial influence of style on the creative process. During a conversation with a co-worker—and he just as baffled—a fact quietly shifts from the insignificant background to the foreground; a trivial anomaly becomes a central piece of evidence, the entire pattern swims into focus, and at last one sees. "How obvious! We knew it all along!" Or a rival may publish first but yet be wrong—and in the crashing wave of fear that he's got it right, followed and engulfed by the wave of realization that it must be wrong, the whole view of the problem skews, the tension of one's concentration twists abruptly higher, and at last one sees. "Not that way, *this* way!"

12 One path to enlightenment, though, has been reported so widely, by writers and artists, by scientists, and especially by mathematicians, that it has become established as a discipline for courting inspiration. The first stage, the reports agree, is prolonged contemplation of the problem, days of saturation in the data, weeks of incessant struggle—the torment of the unknown. The aim is to set in motion the unconscious processes of the mind, to prepare for the intuitive leap. William Lipscomb, a physical chemist at Harvard who won a Nobel Prize for finding the unexpected structures of some unusual molecules, the boranes, said recently that, for him, "The unconscious

mind pieces together random impressions into a continuous story. If I really want to work on a problem, I do a good deal of the work at night—because then I worry about it as I go to sleep." The worry must be about the problem intensely and exclusively. Thought must be free of distraction or competing anxieties. Identification with the problem grows so intimate that the scientist has the experience of the detective who begins to think like the terrorist, of the hunter who feels, as though directly, the silken ripple of the tiger's instincts. One great physical chemist was credited by his peers, who watched him awestruck, with the ability to think about chemical structures directly in quantum terms—so that if a proposed molecular model was too tightly packed he felt uncomfortable, as though his shoes pinched. Joshua Lederberg, president of the Rockefeller University, who won his Nobel for discoveries that established the genetics of microorganisms, said recently, "One needs the ability to strip to the essential attributes of some actor in a process, the ability to imagine oneself *inside* a biological situation; I literally had to be able to think, for example, 'What would it be like if I were one of the chemical pieces in a bacterial chromosome?'—and to try to understand what my environment was, try to know *where* I was, try to know when I was supposed to function in a certain way, and so forth." Total preoccupation to the point of absent-mindedness is no eccentricity—just as the monstrous egoism and contentiousness of some scientists, like that of some artists, are the overflow of the strength and reserves of sureness they must find how they can.

13 Sometimes out of that saturation the answer arises, spontaneous and entire, as though of its own volition. In a famous story, Friedrich Kekulé, a German chemist of the mid-nineteenth century, described how a series of discoveries came to him in the course of hypnagogic reveries—waking dreams. His account, though far from typical, is charming. Kekulé was immersed in one of the most perplexing problems of his day, to find the structural basis of organic chemistry—that is, of the chemistry of compounds that contain carbon atoms. Enormous numbers of such compounds were coming to be known, but their makeup—from atoms of carbon, hydrogen, oxygen, and a few other elements—seemed to follow no rules. Kekulé had dwelt on the compounds' behavior so intensely that the atoms on occasion seemed to appear to him and dance. In the dusk of a summer evening, he was going home by horse-drawn omnibus, sitting outside and alone. "I fell

into a reverie, and lo! The atoms were gamboling before my eyes," he later wrote. "I saw how, frequently, two smaller atoms united to form a pair; how a larger one embraced two smaller ones; how still larger ones kept hold of three or even four of the smaller; whilst the whole kept whirling in a giddy dance. I saw how the larger ones formed a chain." He spent hours that night sketching the forms he had envisioned. Another time, when Kekulé was nodding in his chair before the fire, the atoms danced for him again—but only the larger ones, this time, in long rows, "all twining and twisting in snakelike motion. But look! What was that? One of the snakes had seized hold of its own tail, and the form whirled mockingly before my eyes." The chains and rings that carbon atoms form with each other are indeed the fundamental structures of organic chemistry.

14 Several other scientists have told me that the fringes of sleep set the problem-sodden mind free to make uninhibited, bizarre, even random connections that may throw up the unexpected answer. One said that the technical trick that led to one of his most admired discoveries—it was about the fundamental molecular nature of genetic mutations—had sprung to mind while he was lying insomniac at three in the morning. Another said he was startled from a deep sleep one night by the fully worked-out answer to a puzzle that had blocked him for weeks—though at breakfast he was no longer able to remember any detail except the jubilant certainty. So the next night he went to sleep with paper and pencil on the bedside table; and when, once again, he awoke with the answer, he was able to seize it.

15 More usually, however, in the classic strategy for achieving enlightenment the weeks of saturation must be followed by a second stage that begins when the problem is deliberately set aside. After several days of silence, the solution wells up. The mathematician Henri Poincaré was unusually introspective about the process of discovery. (He also came nearer than anyone else to beating Einstein to the theory of relativity, except that in that case, though he had the pieces of the problem, inspiration did not strike.) In 1908, Poincaré gave a lecture, before the Psychological Society of Paris, about the psychology of mathematical invention, and there he described how he made some of his youthful discoveries. He reassured his audience, few of them mathematical: "I will tell you that I found the proof of a certain theorem in certain circumstances. The theorem will have a barbarous name, which many of you will never have heard of. But that's of no impor-

tance, for what is interesting to the psychologist is not the theorem—it's the circumstances."

16 The youthful discovery was about a class of mathematical functions which he named in honor of another mathematician, Lazarus Fuchs—but, as he said, the mathematical content is not important here. The young Poincaré believed, and for fifteen days he strove to prove, that no functions of the type he was pondering could exist in mathematics. He struggled with the disproof for hours every day. One evening, he happened to drink some black coffee, and couldn't sleep. Like Kekulé with his carbon atoms, Poincaré found mathematical expressions arising before him in crowds, combining and recombining. By the next morning, he had established a class of the functions that he had begun by denying. Then, a short time later, he left town to go on a geological excursion for several days. "The changes of travel made me forget my mathematical work." One day during the excursion, though, he was carrying on a conversation as he was about to board a bus. "At the moment when I put my foot on the step, the idea came to me, without anything in my former thoughts seeming to have paved the way for it, that the transformations I had used to define the Fuchsian functions were identical with those of non-Euclidian geometry." He did not try to prove the idea, but went right on with his conversation. "But I felt a perfect certainty," he wrote. When he got home, "for conscience's sake I verified the result at my leisure."

17 The quality of such moments of the mind has not often been described successfully; Charles P. Snow was a scientist as well as a novelist, and whenever his experience of science comes together with his writer's imagination his witness is authentic. In *The Search*, a novel about scientists at work, the protagonist makes a discovery for which he had long been striving.

> Then I was carried beyond pleasure. . . . My own triumph and delight and success were there, but they seemed insignificant beside this tranquil ecstasy. It was as though I had looked for a truth outside myself, and finding it had become for a moment a part of the truth I sought; as though all the world, the atoms and the stars, were wonderfully clear and close to me, and I to them, so that we were part of a lucidity more tremendous than any mystery.
>
> I had never known that such a moment could exist. . . . Since then I have never quite regained it. But one effect will stay with me as long as I live; once, when I was young, I used to sneer at the mystics

> who have described the experience of being at one with God and part of the unity of things. After that afternoon, I did not want to laugh again; for though I should have interpreted the experience differently, I thought I knew what they meant.

This experience beyond pleasure, like the dark night of the soul, has a name: the novelist Romain Rolland, in a letter to Sigmund Freud, called it "the oceanic sense of well-being."

18 Science is our century's art. Nearly 400 years ago, when modern science was just beginning, Francis Bacon wrote that "knowledge is power." Yet Bacon was not a scientist. He wrote as a bureaucrat in retirement. His slogan was actually the first clear statement of the promise by which, ever since, bureaucrats justify to each other and to king or taxpayer the spending of money on science. Knowledge is power: today we would say, less grandly, that science is essential to technology. Bacon's promise has been fulfilled abundantly, magnificently. The rage to know has been matched by the rage to make. Therefore—with the proviso, abundantly demonstrated, that it's rarely possible to predict which program of fundamental research will produce just what technology and when—the promise has brought scientists in the Western world unprecedented freedom of inquiry. Nonetheless, Bacon's promise hardly penetrates to the thing that moves most scientists. Science has several rewards, but the greatest is that it is the most interesting, difficult, pitiless, exciting, and beautiful pursuit that we have yet found. Science is our century's art.

19 The takeover can be dated more precisely than the beginning of most eras: Friday, June 30, 1905, will do, the day when Albert Einstein, a clerk in the Swiss patent office in Bern, submitted a thirty-one-page paper, "On the Electrodynamics of Moving Bodies," to the journal *Annalen der Physik.* No poem, no play, no piece of music written since then comes near the theory of relativity in its power, as one strains to apprehend it, to make the mind tremble with delight. Whereas fifty years ago it was often said that hardly two score people understood the theory of relativity, today its essential vision, as Einstein himself said, is within reach of any reasonably bright high school student—and that, too, is characteristic of the speed of assimilation of the new in the arts.

20 Consider also the molecular structure of that stuff of the gene,

the celebrated double helix of deoxyribonucleic acid. This is two repetitive strands, one winding up, the other down, but hooked together, across the tube of space between them, by a sequence of pairs of chemical entities—just four sorts of these entities, making just two kinds of pairs, with exactly ten pairs to a full turn of the helix. It's a piece of sculpture. But observe how form and function are one. That sequence possesses a unique duality: one way, it allows the strands to part and each to assemble on itself, by the pairing rules, a duplicate of the complementary strand; the other way, the sequence enciphers, in a four-letter alphabet, the entire specification for the substance of the organism. The structure thus encompasses both heredity and embryological growth, the passing-on of potential and its expression. The structure's elucidation, in March of 1953, was an event of such surpassing explanatory power that it will reverberate through whatever time mankind has remaining. The structure is also perfectly economical and splendidly elegant. There is no sculpture made in this century that is so entrancing.

21 If to compare science to art seems—in the last quarter of this century—to undervalue what science does, that must be, at least partly, because we now expect art to do so little. Before our century, everyone naturally supposed that the artist imitates nature. Aristotle had said so; the idea was obvious, it had flourished and evolved for 2000 years; those who thought about it added that the artist imitates not just nature as it accidentally happens but as it has to be. Yet today that describes the scientist. "Scientific reasoning," Medawar also said, "is a constant interplay or interaction between hypotheses and the logical expectations they give rise to: there is a restless to-and-fro motion of thought, the formulation and reformulation of hypotheses, until we arrive at a hypothesis which, to the best of our prevailing knowledge, will satisfactorily meet the case." Thus far, change only the term "hypothesis" and Medawar described well the experience the painter or the poet has of his own work. "Scientific reasoning is a kind of dialogue between the possible and the actual, between what might be and what is in fact the case," he went on—and there the difference lies. The scientist enjoys the harsher discipline of what is and is not the case. It is he, rather than the painter or the poet in this century, who pursues in its stringent form the imitation of nature.

22 Many scientists—mathematicians and physicists especially—hold that beauty in a theory is itself almost a form of proof. They

speak, for example, of "elegance." Paul Dirac predicted the existence of antimatter (what would science fiction be without him?) several years before any form of it was observed. He won a share in the Nobel Prize in physics in 1933 for the work that included that prediction. "It is more important to have beauty in one's equations than to have them fit experiments," Dirac wrote many years later. "It seems that if one is working from the point of view of getting beauty in one's equations, and if one has really a sound insight, one is on a sure line of progress."

23 Here the scientist parts company with the artist. The insight must be sound. The dialogue is between what might be and what is in fact the case. The scientist is trying to get the thing right. The world is there.

24 And so are other scientists. The social system of science begins with the apprenticeship of the graduate student with a group of his peers and elders in the laboratory of a senior scientist; it continues to collaboration at the bench or the blackboard, and on to formal publication—which is a formal invitation to criticism. The most fundamental function of the social system of science is to enlarge the interplay between imagination and judgment from a private into a public activity. The oceanic feeling of well-being, the true touchstone of the artist, is for the scientist, even the most fortunate and gifted, only the midpoint of the process of doing science.

Suggestions for Discussion and Writing

1. Judson describes the "moments of truth" that inspire many scientists. Do you think such moments are limited to scientists? What other kinds of people might know such moments?

2. The purpose of Judson's essay is obviously to tell readers something about the rage to know. But what is his thesis? Where in the essay is it expressed? Do you think the essay proves his thesis?

3. Judson explains in great detail the process by which scientific discoveries are made. It sounds very much like the process many writers use to create a text. Write down all the steps you take in writing a paper; how "scientific" is your writing process? How different is your process from those of your classmates?

4. Assume that you have been asked to participate in a campus debate on "Highlights of the Twentieth Century." By luck of the draw, you are assigned to debate Judson's claim that science took over art on June 30, 1905, when Albert Einstein submitted his special theory of relativity for publication. Would you agree with him? What evidence, pro or con, would you use to support your argument? (Assume that your audience will be fellow students and some faculty members.)

5. Many of the scientists Judson quotes stress the spontaneity and enjoyment of doing science. How closely does this match your idea of what science should be like? Why do you think Rosalind Franklin didn't want the funding agencies to know "that (science is) so much fun"? Does this say something about our culture's definition of science?

J. Robert Oppenheimer

Prospects in the Arts and Sciences

JULIUS ROBERT OPPENHEIMER was born in New York in 1904. Known as the "father of the atomic bomb," he served as director of the Atomic Energy Lab at Los Alamos from 1946-1952, and chaired the UN Atomic Energy Commission from 1947-1952. His work with advanced weapons led him to voice doubt about the development of the hydrogen bomb, fearing it would imperil world peace. In the Eisenhower/McCarthy years, these were dangerous sentiments; Oppenheimer found himself accused of being a Communist and classified as a security risk to the United States. Although he was never proven disloyal, he had to leave public service; he became director of Princeton's Institute for Advanced Study, and served there until his death in 1967.

In a speech Oppenheimer said "It is not possible to be a scientist unless you believe that it is good to learn. . . . It is not possible to be a scientist unless you believe that the knowledge of the world, and the power which this gives, is a thing of intrinsic value to humanity, and that you are using it to help in the spread of knowledge, and are willing to take the consequences." The speech below was given to a symposium at Columbia University in 1954.

In the natural sciences these are, and have been, and are most surely likely to continue to be, heroic days. Discovery follows discovery, each both raising and answering questions, each ending a long search, and each providing the new instruments for new search.

2 There are radical ways of thinking unfamiliar to common sense, connected with it by decades or centuries of increasingly specialized and unfamiliar experience. There are lessons how limited, for all its variety, the common experience of man has been with regard to natural phenomenon, and hints and analogies as to how limited may be his experience with man.

3 Every new finding is a part of the instrument kit of the sciences for further investigation and for penetrating into new fields. Discoveries of knowledge fructify technology and the practical arts, and these in turn pay back refined techniques, new possibilities for observation and experiment.

4 In any science there is a harmony between practitioners. A man may work as an individual, learning of what his colleagues do through reading or conversation; or he may be working as a member of a group on problems whose technical equipment is too massive for individual effort. But whether he is part of a team or solitary in his own study, he, as a professional, is a member of a community.

5 His colleagues in his own branch of science will be grateful to him for the inventive or creative thoughts he has, will welcome his criticism. His world and work will be objectively communicable and he will be quite sure that, if there is error in it, that error will not be long undetected. In his own line of work he lives in a community where common understanding combines with common purpose and interest to bind men together both in freedom and in cooperation.

6 This experience will make him acutely aware of how limited, how precious is this condition of his life; for in his relations with a wider society there will be neither the sense of community nor of objective understanding. He will sometimes find, it is true, in return-

ing to practical undertakings, some sense of community with men who are not expert in his science, with other scientists whose work is remote from his, and with men of action and men of art.

7 The frontiers of science are separated now by long years of study, by specialized vocabularies, arts, techniques and knowledge from the common heritage even of a most civilized society, and anyone working at the frontier of such science is in that sense a very long way from home and a long way, too, from the practical arts that were its matrix and origin, as indeed they were of what we today call art.

8 The specialization of science is an inevitable accompaniment of progress; yet it is full of dangers, and it is cruelly wasteful, since so much that is beautiful and enlightening is cut off from most of the world. Thus it is proper to the role of the scientist that he not merely find new truth and communicate it to his fellows, but that he teach, that he try to bring the most honest and intelligible account of new knowledge to all who will try to learn.

9 This is one reason—it is the decisive organic reason—why scientists belong in universities. It is one reason why the patronage of science by and through universities is its most proper form; for it is here, in teaching, in the association of scholars, and in the friendships of teachers and taught, of men who by profession must themselves be both teachers and taught, that the narrowness of scientific life can best be moderated and that the analogies, insights and harmonies of scientific discovery can find their way into the wider life of man.

10 In the situation of the artist today there are both analogies and differences to that of the scientist; but it is the differences which are the most striking and which raise the problems that touch most on the evil of our day.

11 For the artist it is not enough that he communicate with others who are expert in his own art. Their fellowship, their understanding and their appreciation may encourage him; but that is not the end of his work, nor its nature.

12 The artist depends on a common sensibility and culture, on a common meaning of symbols, on a community of experience and common ways of describing and interpreting it. He need not write for everyone or paint or play for everyone. But his audience must be man, and not a specialized set of experts among his fellows.

13 Today that is very difficult. Often the artist has an aching sense of great loneliness, for the community to which he addresses himself

is largely not there; the traditions and the history, the myths and the common experience, which it is his function to illuminate and to harmonize and to portray, have been dissolved in a changing world.

14 There is, it is true, an artificial audience maintained to moderate between the artist and the world for which he works: the audience of the professional critics, popularizers and advertisers of art. But though, as does the popularizer and promoter of science, the critic fulfills a necessary present function, and introduces some order and some communication between the artist and the world, he cannot add to the intimacy and the directness and the depth with which the artist addresses his fellow men.

15 To the artist's loneliness there is a complementary great and terrible barrenness in the lives of men. They are deprived of the illumination, the light and the tenderness and insight of an intelligible interpretation, in contemporary terms, of the sorrows and wonders and gaeties and follies of man's life.

16 This may be in part offset, and is, by the great growth of technical means for making the art of the past available. But these provide a record of past intimacies between art and life; even when they are applied to the writing and painting and composing of the day, they do not bridge the gulf between a society too vast and too disordered and the artist trying to give meaning and beauty to its parts.

17 In an important sense, this world of ours is a new world, in which the unity of knowledge, the nature of human communities, the order of society, the order of ideas, the very notions of society and culture have changed and will not return to what they have been in the past. What is new is new not because it has never been there before, but because it has changed in quality.

18 One thing that is new is the prevalence of newness, the changing scale and scope of change itself, so that the world alters as we walk in it, so that the years of man's life measure not some small growth or rearrangement or moderation of what he learned in childhood, but a great upheaval.

19 What is new is that in one generation our knowledge of the natural world engulfs, upsets and complements all knowledge of the natural world before. The techniques, among which and by which we live, multiply and ramify, so that the whole world is bound together by communication, blocked here and there by the immense synopses of political tyranny.

20 The global quality of the world is new: our knowledge of and sympathy with remote and diverse peoples, our involvement with them in practical terms and our commitment to them in terms of brotherhood. What is new in the world is the massive character of the dissolution and corruption of authority, in belief, in ritual and in temporal order.

21 Yet this is the world that we have come to live in. The very difficulties which it presents derive from growth in understanding, in skill, in power. To assail the changes that have unmoored us from the past is futile, and, in a deep sense, I think it is wicked. We need to recognize the change and learn what resources we have.

22 Again I will turn to the schools, and, as their end and as their center, the universities. For the problem of the scientist is in this respect not different from that of the artist, nor of the historian. He needs to be a part of the community, and the community can only, with loss and peril, be without him. Thus it is with a sense of interest and hope that we see a growing recognition that the creative artist is a proper charge on the university, and the university a proper home for him: that a composer or a poet or a playwright or painter needs the toleration, understanding, the rather local and parochial patronage that a university can give; and that this will protect him to some extent from the tyranny of man's communication and professional promotion.

23 For here there is an honest chance that what the artist has of insight and of beauty will take root in the community and that some intimacy and some human bonds can mark his relations with his patrons. For a university rightly and inherently is a place where the individual man can form new syntheses, where the accidents of friendship and association can open a man's eyes to a part of science or art which he had not known before, where parts of human life, remote and perhaps superficially incompatible one with the other, can find in men their harmony and their synthesis.

24 The truth is that this is indeed inevitably and increasingly an open, and inevitably and increasing an eclectic world. We know too much for one man to know much, we live too variously to live as one. Our histories and traditions—the very means of interpreting life—are both bonds and barriers among us. Our knowledge separates as well as it unites; our orders disintegrate as well as bind; our art brings us together and sets us apart. The artist's loneliness, the scholar's despairing, because no one will any longer trouble to learn what he can

teach, the narrowness of the scientist, these are not unnatural insignia in this great time of change.

25 This is a world in which each of us, knowing his limitations, knowing the evils of superficiality and the terrors of fatigue, will have to cling to what is close to him, to what he knows, to what he can do, to his friends and his tradition and his love, lest he be dissolved in a universal confusion and know nothing and love nothing.

26 Both the man of science and the man of art live always at the edge of mystery, surrounded by it; both always, as the measure of their creation, have had to do with the harmonization of what is new and what is familiar, with the balance between novelty and synthesis, with the struggle to make partial order in total chaos.

27 This cannot be an easy life. We shall have a rugged time of it to keep our minds open and to keep them deep, to keep our sense of beauty and our ability to make it, and our occasional ability to see it, in places remote and strange and unfamiliar.

28 But this is, as I see it, the condition of man; and in this condition we can help, because we can love one another.

Suggestions for Discussion and Writing

1. Oppenheimer's title uses a familiar pair of opposites, "arts" and "sciences". Do you think he sees these as distinct areas? Why do we make this distinction in our culture?

2. This essay offers a strong argument that science belongs in universities. Where else will you find scientists working today? What would Oppenheimer have to say about locating science in these communities?

3. At the time Oppenheimer made this argument, it was considered "unAmerican" to argue that scientific discoveries should be shared by everyone, and not "belong" to individual nations. Has that attitude changed in the forty years since he made this speech? In what ways?

4. When Oppenheimer refers to an individual in this essay, he uses the terms "man" or "he". What effect do these terms have on contemporary readers? Do they suggest that only men should be scientists? How could you revise his language for a modern audience?

5. Oppenheimer's language and examples seem carefully chosen for his audience and his political situation. Pick out examples that you think were particularly suited to a speech given at the height of the Cold War. How do Oppenheimer's choices affect readers of his essay? What picture of him do they convey?

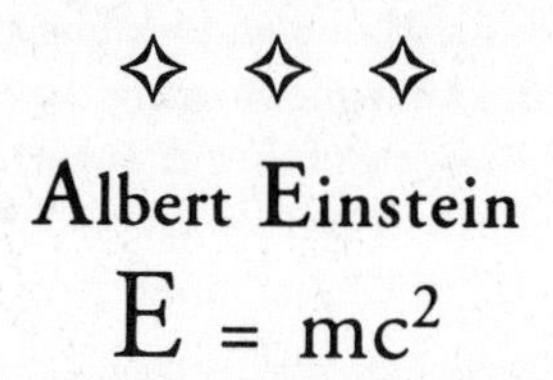

Albert Einstein

$E = mc^2$

Born in Germany in 1879, ALBERT EINSTEIN was educated in Switzerland, where as most of us know, he had great trouble with his math classes. He received his Ph.D. in physics from the University of Zurich in 1905. In that same year Einstein proposed his special theory of relativity; by 1915 he had expanded this into his theory of general relativity, which is explained in this essay. He was awarded the Nobel Prize in Physics in 1921.

Einstein was an avowed pacifist and a spokesperson for Zionism. Pressures to curtail his political activities led him to leave Germany in 1933, shortly before Hitler rose to power; he settled in Princeton, New Jersey. While he continued intense scientific investigations, he remained active as a pacifist, campaigning for the abolition of war and the controlled development of nuclear weapons to ensure the safety of humanity. In 1952 he refused the presidency of Israel. He told another great mathematician, Stephen Hawking, "Equations are more important to me. Politics is for the present, but an equation is something for eternity." He died in 1955. This essay on general relativity was published in *Science Illustrated*, a magazine for laypeople interested in science, in 1946.

In order to understand the law of the equivalence of mass and energy, we must go back to two conservation or "balance" principles which, independent of each other, held a high place in pre-relativity physics. These were the principle of the conservation of energy and the principle of the conservation of mass. The first of these, advanced by

Leibnitz as long ago as the seventeenth century, was developed in the nineteenth century essentially as a corollary of a principle of mechanics.

2 Consider, for example, a pendulum whose mass swings back and forth between the points A and B. At these points the mass m is higher by the amount h than it is at C, the lowest point of the path (see drawing). At C, on the other hand, the lifting height has disappeared and instead of it the mass has a velocity v. It is as though the lifting height could be converted entirely into velocity, and vice versa. The exact relation would be expressed as

$$mgh = \frac{m}{2}v^2$$

with g representing the acceleration of gravity. What is interesting here is that this relation is independent of both the length of the pendulum and the form of the path through which the mass moves.

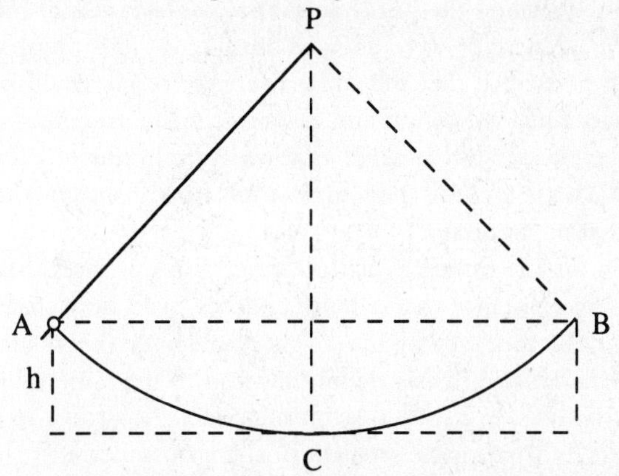

3 The significance is that something remains constant throughout the process, and that something is energy. At A and at B it is an energy of position, or "potential" energy; at C it is an energy of motion, or "kinetic" energy. If this concept is correct, then the sum

$$mgh + m\frac{v^2}{2}$$

must have the same value for any position of the pendulum, if h is understood to represent the height above C, and v the velocity at that point in the pendulum's path. And such is found to be actually the case. The generalization of this principle gives us the law of the conservation of mechanical energy. But what happens when friction stops the pendulum?

4 The answer to that was found in the study of heat phenomena. This study, based on the assumption that heat is an indestructible substance which flows from a warmer to a colder object, seemed to give us a principle of the "conservation of heat." On the other hand, from time immemorial it has been known that heat could be produced by friction, as in the fire-making drills of the Indians. The physicists were for long unable to account for this kind of heat "production." Their difficulties were overcome only when it was successfully established that, for any given amount of heat produced by friction, an exactly proportional amount of energy had to be expended. Thus did we arrive at a principle of the "equivalence of work and heat." With our pendulum, for example, mechanical energy is gradually converted by friction into heat.

5 In such fashion the principles of the conservation of mechanical and thermal energies were merged into one. The physicists were thereupon persuaded that the conservation principle could be further extended to take in chemical and electromagnetic processes—in short, could be applied to all fields. It appeared that in our physical system there was a sum total of energies that remained constant through all changes that might occur.

6 Now for the principle of the conservation of mass. Mass is defined by the resistance that a body opposes to its acceleration (inert mass). It is also measured by the weight of the body (heavy mass). That these two radically different definitions lead to the same value for the mass of a body is, in itself, an astonishing fact. According to the principle—namely, that masses remain unchanged under any physical or chemical changes—the mass appeared to be the essential (because unvarying) quality of matter. Heating, melting, vaporization, or combining into chemical compounds would not change the total mass.

7 Physicists accepted this principle up to a few decades ago. But it proved inadequate in the face of the special theory of relativity. It was therefore merged with the energy principle—just as, about 60 years before, the principle of the conservation of mechanical energy had been combined with the principle of the conservation of heat. We might say that the principle of the conservation of energy, having previously swallowed up that of the conservation of heat, now proceeded to swallow that of the conservation of mass—and holds the field alone.

8 It is customary to express the equivalence of mass and energy (though somewhat inexactly) by the formula $E = mc^2$, in which c represents the velocity of light, about 186,000 miles per second. E is the energy that is contained in a stationary body; m is its mass. The energy that belongs to the mass m is equal to this mass, multiplied by the square of the enormous speed of light—which is to say, a vast amount of energy for every unit of mass.

9 But if every gram of material contains this tremendous energy, why did it go so long unnoticed? The answer is simple enough: so long as none of the energy is given off externally, it cannot be observed. It is as though a man who is fabulously rich should never spend or give away a cent; no one could tell how rich he was.

10 Now we can reverse the relation and say that an increase of E in the amount of energy must be accompanied by an increase of $\frac{E}{c^2}$ in the mass. I can easily supply energy to the mass—for instance, if I heat it by 10 degrees. So why not measure the mass increase, or weight increase, connected with this change? The trouble here is that in the mass increase the enormous factor c^2 occurs in the denominator of the fraction. In such a case the increase is too small to be measured directly; even with the most sensitive balance.

11 For a mass increase to be measurable, the change of energy per mass unit must be enormously large. We know of only one sphere in which such amounts of energy per mass unit are released: namely, radioactive disintegration. Schematically, the process goes like this: an atom of the mass M splits into two atoms of the mass M' and M", which separate with tremendous kinetic energy. If we imagine these two masses as brought to rest—that is, if we take this energy of motion from them—then, considered together, they are essentially poorer in energy than was the original atom. According to the equivalence principle, the mass sum M' + M" of the disintegration products must also be somewhat smaller than the original mass M of the disintegrating atom—in contradiction to the old principle of the conservation of mass. The relative difference of the two is on the order of $\frac{1}{10}$ of one percent.

12 Now, we cannot actually weigh the atoms individually. However, there are indirect methods for measuring their weights exactly. We can likewise determine the kinetic energies that are transferred to the disintegration products M' and M". Thus it has become

possible to test and confirm the equivalence formula. Also, the law permits us to calculate in advance, from precisely determined atom weights, just how much energy will be released with any atom disintegration we have in mind. The law says nothing, of course, as to whether—or how—the disintegration reaction can be brought about.

13 What takes place can be illustrated with the help of our rich man. The atom M is a rich miser who, during his life, gives away no money *(energy)*. But in his will he bequeaths his fortune to his sons M' and M", on condition that they give to the community a small amount, less than one thousandth of the whole estate *(energy or mass)*. The sons together have somewhat less than the father had *(the mass sum M' + M" is somewhat smaller than the mass M of the radioactive atom)*. But the part given to the community, though relatively small, is still so enormously large *(considered as kinetic energy)* that it brings with it a great threat of evil. Averting that threat has become the most urgent problem of our time.

Suggestions for Discussion and Writing

1. Would Brownowski agree with Einstein that physics should have an interest for the general public? Do you agree? Why or why not?

2. Relativity is not a simple concept, as people who have struggled with it in physics classes can attest. Yet in this essay, Einstein manages to make it quite clear. Identify as many techniques as you can that he uses to explain this concept. Which seem most effective to you?

3. What scientific terms does Einstein define in the essay? Which ones doesn't he define? What assumptions about his readers do these definitions suggest?

4. What is the purpose of Einstein's conclusion? Were you surprised by the last sentence? Reread the essay; can you find any clues "setting up" this last sentence? Why do you think he included it?

5. What role did Einstein, an avowed pacifist, play in the development of atomic and nuclear weapons? From what your research tells you about that role, write a brief analysis speculating why Einstein wrote this essay.

2 Science vs. Pseudoscience

People like to have understandable explanations for phenomena in the world around them. When they don't understand scientific principles fully, they are likely to evolve complex "pseudoscientific" explanations that account for things they don't understand rationally. For instance, most cultures have creation myths, and until a few centuries ago, we believed that the earth was flat and that babies grew from tiny, fully-formed bodies called *homunculi* inside the womb. Even today, we can find pseudoscience all around us. A popular television ad for antacid, for instance, tells us to prefer the brand containing calcium to those containing aluminum and magnesium, because our bodies need calcium. Well, we need aluminum and magnesium, too; and the amounts and kinds of those metals in any antacid product aren't likely to be absorbed by the body in the first place. However, because we pseudo-scientifically believe that only calcium is important for our bodies, the manufacturer can make these claims.

In this section we provide readings that examine some of the most common pseudo-scientific beliefs of our time. Niles Eldredge tackles "creation science" on behalf of evolutionists; then Tom Bethell lets us see that evolution has other challengers. Astronomer Bart J. Bok explains why astrology isn't a science, and Alfred Meyer looks at what "scientific" machines like lie detectors can really detect. As you read these essays, ask yourself, "What kinds of scientific claims am I asked to believe each day? What evidence do the people who want me to believe these claims give me? Must everything be physically verifiable? Can science co-exist with faith?"

Niles Eldredge

Creationism Isn't Science

NILES ELDREDGE is a paleontologist and curator of the Department of Invertebrates at the American Museum of Natural History in New York. With Stephen Jay Gould he is the leading proponent of "punctuated equilibrium," the theory that evolution happens suddenly when a small branch of a species undergoes rapid change, instead of gradually across an entire species. He is the author of a number of books, including *Unfinished Synthesis* (1985), *Time Frames: The Rethinking of Darwinian Evolution* (1986), *Life Pulse: Episodes from the Story of the Fossil Record* (1987), and *The Fossil Factory* (1989). His most recent book is *The Miner's Canary: A Paleontologist Unravels the Mysteries of Extinction* (1991). This essay appeared in *The New Republic* in 1981.

Despite this country's apparent modernism, the creationist movement once again is growing. The news media proclaimed a juryless trial in California as "Scopes II" and those who cling to the myth of progress wonder how the country could revert to the primitive state it was in when Darrow and Bryan battled it out in the hot summer of 1925 in Dayton, Tennessee. But the sad truth is that we have not progressed. Creationism never completely disappeared as a political, religious, and educational issue. Scopes was convicted of violating the Tennessee statute forbidding the teaching of the evolutionary origins of mankind (although in fact he was ill and never really did teach the evolution segment of the curriculum). The result was a drastic cutback in serious discussion of evolution in many high school texts until it became respectable again in the 1960s.

2 Although technological advances since 1925 have been prodigious, and although science news magazines are springing up like toadstools, the American public appears to be as badly informed about the real nature of science as it ever was. Such undiluted ignorance, coupled with the strong anti-intellectual tradition in the U.S., provides a congenial climate for creationism to leap once more to the fore, along with school prayer, sex education, Proposition 13, and the other favorite issues of the populist, conservative movement. Much of the success of recent creationist efforts lies in a prior failure to educate our children about science—how it is done, by whom, and how its results are to be interpreted.

3 Today's creationists usually cry for "equal time" rather than for actually substituting the Genesis version of the origin of things for the explanations preferred by modern science. (The recent trial in California is an anachronism in this respect because the plaintiff simply affirmed that his rights of religious freedom were abrogated by teaching him that man "descended from apes".) At the heart of the creationists' contemporary political argument is an appeal to the time-honored American sense of fair play. "Look," they say, "evolution is only a theory. Scientists cannot agree on all details either of the exact course of evolutionary history, or how evolution actually takes place." True enough. Creationists then declare that many scientists have grave doubts that evolution actually has occurred—a charge echoed by Ronald Reagan during the campaign, and definitely false. They argue that since evolution is only a theory, why not, in the spirit of fair play, give equal time to equally plausible explanations of the origin of the cosmos, of life on earth, and of mankind? Why not indeed?

4 The creationist argument equates a biological, evolutionary system with a non-scientific system of explaining life's exuberant diversity. Both systems are presented as authoritarian, and here lies the real tragedy of American science education: the public is depressingly willing to see merit in the "fair play, equal time" argument precisely because it views science almost wholly in this authoritarian vein. The public is bombarded with a constant stream of oracular pronounce-

ments of new discoveries, new truths, and medical and technological innovations, but the American education system gives most people no effective choice but to ignore, accept on faith, or reject out of hand each n`ew scientific finding. Scientists themselves promote an Olympian status for their profession; it's small wonder that the public has a tough time deciding which set of authoritarian pronouncements to heed. So why not present them all and let each person choose his or her own set of beliefs?

5 Of course, there has to be some willingness to accept the expertise of specialists. Although most of us "believe" the earth is spherical, how many of us can design and perform an experiment to show that it must be so? But to stress the authoritarianism of science is to miss its essence. Science is the enterprise of comparing alternative ideas about what the cosmos is, how it works, and how it came to be. Some ideas are better than others, and the criterion for judging which are better is simply the relative power of different ideas to fit our observations. The goal is greater understanding of the natural universe. The method consists of constantly challenging received ideas, modifying them, or, best of all, replacing them with better ones.

6 So science is ideas, and the ideas are acknowledged to be merely approximations to the truth. Nothing could be further from authoritarianism—dogmatic assertions of what is true. Scientists deal with ideas that appear to be the best (the closest to the truth) given what they think they know about the universe at any given moment. If scientists frequently act as if their ideas *are* the truth, they are simply showing their humanity. But the human quest for a rational coming-to-grips with the cosmos recognizes imperfection in observation and thought, and incorporates the frailty into its method. Creationists disdain this quest, preferring the wholly authoritarian, allegedly "revealed" truth of divine creation as an understanding of our beginnings. At the same time they present disagreement among scientists as an expression of scientific failure in the realm of evolutionary biology.

7 To the charge that "evolution is *only* a theory," we say "all science is theory." Theories are ideas, or complex sets of ideas, which explain some aspect of the natural world. Competing theories sometimes coexist until one drives the other out, or until both are discarded in favor of yet another theory. But it is true that one major theory usually holds sway at any one time. All biologists, including biochemists, molecular geneticists, physiologists, behaviorists, and anatomists, see a

pattern of similarity interlocking the spectrum of millions of species, from bacteria to timber wolves. Darwin finally convinced the world that this pattern of similarity is neatly explained by "descent with modification." If we imagine a genealogical system where an ancestor produces one or more descendants, we get a pattern of progressive similarity. The whole array of ancestors and descendants will share some feature inherited from the first ancestor; as each novelty appears, it is shared only with later descendants. All forms of life have the nucleic acid RNA. One major branch of life, the vertebrates, all share backbones. All mammals have three inner ear bones, hair, and mammary glands. All dogs share features not found in other carnivores, such as cats. In other words, dogs share similarities among themselves in addition to having general mammalian features, plus general vertebrate features, as well as anatomical and biochemical similarities shared with the rest of life.

8 How do we test the basic notion that life has evolved? The notion of evolution, like any scientific idea, should generate predictions about the natural world, which we can discover to be true or false. The grand prediction of evolution is that there should be one basic scheme of similarities interlocking all of life. This is what we have consistently found for over 100 years, as thousands of biologists daily compared different organisms. Medical experimentation depends upon the interrelatedness of life. We test drugs on rhesus monkeys and study the effects of caffeine on rats because we cannot experiment on ourselves. The physiological systems of monkeys are more similar to our own than to rats. Medical scientists know this and rely on this prediction to interpret the significance of their results in terms of human medicine. Very simply, were life not all interrelated, none of this would be possible. There would be chaos, not order, in the natural world. There is no competing, rational biological explanation for this order in nature, and there hasn't been for a long while.

9 Creationists, of course, have an alternative explanation for this order permeating life's diversity. It is simply the way the supernatural creator chose to pattern life. But any possible pattern could be there, including chaos—an absence of any similarity among the "kinds" of organisms on earth—and creationism would hold that it is just what the creator made. There is nothing about this view of life that smacks of prediction. It tells us nothing about what to expect if we begin to study organisms in detail. In short, there is nothing in this notion that

allows us to go to nature to test it, to verify or reject it.

10 And there is the key difference. Creationism (and it comes in many guises, most of which do not stem from the Judeo-Christian tradition) is a belief system involving the supernatural. Testing an idea with our own experiences in the natural universe is simply out of bounds. The mystical revelation behind creationism is the opposite of science, which seeks rational understanding of the cosmos. Science thrives on alternative explanations, which must be equally subject to observational and experimental testing. No form of creationism even remotely qualifies for inclusion in a science curriculum.

11 Creationists have introduced equal-time bills in over 10 state legislatures, and recently met with success when Governor White of Arkansas signed such a bill into law on March 19 (reportedly without reading it). Creationists also have lobbied extensively at local school boards. The impact has been enormous. Just as the latest creationist bill is defeated in committee, and some of their more able spokesmen look silly on national TV, one hears of a local school district in the Philadelphia environs where some of the teachers have adopted the "equal time" or "dual model" approach to discussing "origins" in the biology curriculum on their own initiative. Each creationist "defeat" amounts to a Pyrrhic victory for their opponents. Increasingly, teachers are left to their own discretion, and whether out of personal conviction, a desire to be "fair," or fear of parental reprisal, they are teaching creationism along with evolution in their biology classes. It is simply the path of least resistance.

12 Acceptance of equal time for two alternative authoritarian explanations is a startling blow to the fabric of science education. The fundamental notion a student should get from high school science is that people can confront the universe and learn about it directly. Just one major inroad against this basic aspect of science threatens all of science education. Chemistry, physics, and geology—all of which spurn biblical revelation in favor of direct experience, as all science must—are jeopardized every bit as much as biology. That some creationists have explicitly attacked areas of geology, chemistry, and physics (in arguments over the age of the earth, for example) underscores the more general threat they pose to all science. We must remove science education from its role as authoritarian truthgiver. This view distorts the real nature of science and gives creationists their most potent argument.

13 The creationists' equal-time appeal maintains that evolution itself amounts to a religious belief (allied with a secular humanism) and should not be included in a science curriculum. But if it is included, goes the argument, it must appear along with other religious notions. Both are authoritarian belief systems, and neither is science, according to this creationist ploy.

14 The more common creationist approach these days avoids such sophistry and maintains that both creationism and evolution belong in the realm of science. But apart from some attempts to document the remains of Noah's Ark on the flanks of Mt. Ararat, creationists have been singularly unsuccessful in posing testable theories about the origin, diversity, and distribution of plants and animals. No such contributions have appeared to date either in creationism's voluminous literature or, more to the point, in the professional biological literature. "Science creationism" consists almost exclusively of a multipronged attack on evolutionary biology and historical geology. No evidence, for example, is presented in favor of the notion that the earth is only 20,000 years old, but many arguments attempt to poke holes in geochemists' and astronomers' reckoning of old Mother Earth's age at about 4.6 billion years. Analysis of the age of formation of rocks is based ultimately on the theories of radioactive decay in nuclear physics. (A body of rock is dated, often by several different means, in several different laboratories. The results consistently agree. And rocks shown to be roughly the same age on independent criteria (usually involving fossils) invariably check out to be roughly the same age when dated radiometrically. The system, although not without its flaws, works.) The supposed vast age of any particular rock can be shown to be false, but not by quoting Scripture.

15 All of the prodigious works of "scientific creationism" are of this nature. All can be refuted. However, before school boards or parent groups, creationists are fond of "debating" scientists by bombarding the typically ill-prepared biologist or geologist with a plethora of allegations, ranging from the second law of thermodynamics (said to falsify evolution outright) to the supposed absence of fossils intermediate between "major kinds." No scientist is equally at home in all realms of physics, chemistry, geology, and biology in this day of advanced specialization. Not all the proper retorts spring readily to mind. Retorts there are, but the game is usually lost anyway, as rebuttals strike an audience as simply another set of authoritarian

statements they must take on faith.

16 Although creationists persist in depicting both science and creationism as two comparable, monolithic belief systems, perhaps the most insidious attack exploits free inquiry in science. Because we disagree on specifics, some of my colleagues and I are said now to have serious doubts that evolution has occurred. Distressing as this may be, the argument actually highlights the core issue raised by creationism. The creationists are acknowledging that science is no monolithic authoritarian belief system. But even though they recognize that there are competing ideas within contemporary biology, the creationists see scientific debate as a sign of weakness. Of course, it really is a sign of vitality.

17 Evolutionary theory since the 1940s (until comparatively recently) has focused on a single coherent view of the evolutionary process. Biologists of all disciplines agree to a remarkable degree on the outlines of this theory, the so-called "modern synthesis." In a nutshell, this was a vindication of Darwin's original position: that evolution is predominantly an affair of gradual progressive change. As environmental conditions changed, natural selection (a culling process similar to the "artificial" selection practiced by animal breeders) favored those variants best suited to the new conditions. Thus evolutionary change is fundamentally adaptive. The modern synthesis integrated the newly arisen science of genetics with the Darwinian view and held that the entire diversity of life could be explained in these simple terms.

18 Some biologists have attacked natural selection itself, but much of the current uproar in evolutionary biology is less radical in implication. Some critics see a greater role for random processes. Others, like me, see little evidence of gradual, progressive change in the fossil record. We maintain that the usual explanation—the inadequacy of the fossil record—is itself inadequate to explain the non-change, the maintenance of status quo which lasts in some cases for 10 million years or more in our fossil bones and shells. In this view, change (presumably by natural selection causing adaptive modifications) takes place in bursts of a few thousand years, followed usually by immensely longer periods of business as usual.

19 Arguments become heated. Charges of "straw man," "no evidence," and so on are flung about—which shows that scientists, like everyone, get their egos wrapped up in their work. They believe passion-

ately in their own ideas, even if they are supposed to be calm, cool, dispassionate, and able to evaluate all possibilities evenly. (It is usually in the collective process of argument that the better ideas win out in science; seldom has anyone single-handedly evinced the open-mindedness necessary to drop a pet idea.) But nowhere in this *sturm und drang* has any of the participants come close to denying that evolution has occurred.

20 So the creationists distort. An attack on some parts of Darwin's views is equated with a rejection of evolution. They conveniently ignore that Darwin merely proposed one of many sets of ideas on *how* evolution works. The only real defense against such tactics lies in a true appreciation of the scientific enterprise—the trial-and-error comparison of ideas and how they seem to fit the material universe. If the public were more aware that scientists are expected to disagree, that what a scientist writes today is not the last word, but a progress report on some very intensive thinking and investigation, creationists would be far less successful in injecting an authoritarian system of belief into curricula supposedly devoted to free, open rational inquiry into the nature of natural things.

Suggestions for Discussion and Writing

1. What is Eldredge's purpose in writing this essay? Does his concern about creationism rise only from the fact that he is an evolutionist? What dangers does creationism pose to science?

2. Regarding the creation vs. evolution dispute, Eldredge's colleague Stephen Jay Gould said, "The reaction of evolutionists to creation is very personal. We are a small profession. Some three thousand people in this country spend their professional lifetimes studying evolution, and creationism is a direct threat to one of the most exciting things scientists have ever learned. Of course we had to fight it. It wasn't public service—though aspects of it turned out to be public service." Do you think Eldredge shares Gould's views? What evidence in the essay leads you to your conclusions?

3. Much of Eldredge's argument revolves around the distinction between *theory* and *fact.* What, in his view, is the difference? How persuaded are you by his distinction? Would Lewis Thomas agree with him?

4. In Eldredge's opinion, what is wrong with the "equal time" argument made by the creationists? Do you agree with him?

5. How is the development of species treated in schools in your community? Do you see evidence that schools are taking "the path of least resistance"? What changes, if any, would you make in natural history education in your community? Present your conclusions as a report to your local school or library board.

Tom Bethell
Agnostic Evolutionists

Born in London in 1940, Tom Bethell came to the U.S. in 1962 after studying at the Royal Naval College. He held a number of jobs, including one as assistant to New Orleans District Attorney Jim Garrison during Garrison's investigation of John Fitzgerald Kennedy's assassination. Then he became Washington editor for *Harper's* magazine; currently he is Washington correspondent and an occasional columnist for *American Spectator*. Concerning religion and evolution, he says, "I believe in Original Sin, and I prefer the genetics of Mendel to that of Lysenko."

A devoted music fan whose tastes run from Gregorian chants to Beethoven to Jelly Roll Morton, Bethell claims to have spent a quarter of his life listening to music. He has also written *George Lewis: A Jazzman from New Orleans* (1977) and an influential analysis of television's coverage of economic news. *The Electric Windmill: An Inadvertent Autobiography* appeared in 1988. One reviewer wrote, "He can parallel park an argument more neatly than anyone else going. . . . I can't imagine any editor rejecting anything submitted by Tom Bethell." This essay was originally published in *Harper's* in 1985.

The first time I saw Colin Patterson was at the American Museum of Natural History in New York City in the spring of 1983. He was in the office of Donn Rosen, a curator in the museum's department of ichthyology, which is the branch of zoology that deals with fishes. Patterson, a paleontologist specializing in fossil fishes, was staring through a binocular microscope at a slice of codfish. In his mid-fifties and balding, he was wearing black corduroys and a smoking-jacket affair of the kind that I associate with the Sloane Square poets of the "angry young man" generation—the generation to which Patterson belongs by age, and perhaps by temperament. I would later spend time with him in London, at the British Museum of Natural History, where he is a senior paleontologist, and at Cambridge University, where we attended a lecture by the famous Harvard paleontologist Stephen Jay Gould. He often conveyed an impression of moody rebelliousness: he is authoritative, the kind of person others defer to in a discussion; he is habitually pessimistic; and he seemed not at all sanguine about his brushes with other scientists—encounters that by the late 1970s had become quite frequent. Those with whom Patterson has been arguing are mostly paleontologists and evolutionary biologists—researchers and academics who have devoted their careers, their lives, to upholding and fine-tuning the ideas about the origins and the development of species introduced by Charles Darwin in the second half of the nineteenth century. Patterson, it seemed, was no longer sure he believed in evolutionary theory, and he was saying so. Or, perhaps more accurately, he was saying that evolutionists—like the creationists they periodically do battle with—are nothing more than believers themselves.

2 In 1978, Patterson wrote an introductory book called *Evolution*, which was published by the British Museum. A year later, he received a letter from Luther Sunderland, an electrical engineer in upstate New York and a creationist-activist, asking why *Evolution* did not include any "direct illustrations of evolutionary transitions." Patterson's reply included the following:

> You say I should at least "show a photo of the fossil from which each type of organism was derived." I will lay it on the line—there is not one such fossil for which one could make a watertight argument. The reason is that statements about ancestry and descent are not applicable in the fossil record. Is *Archaeopteryx* the ancestor of all birds? Perhaps yes, perhaps no: there is no way of answering the question. It is easy enough

> to make up stories of how one form gave rise to another, and to find reasons why the stages should be favoured by natural selection. But such stories are not part of science, for there is no way of putting them to the test.

3 By 1981, Patterson's doubts about evolutionary theory were finding their way to the public. A sentence in a brochure he wrote that year for the British Museum began: "If the theory of evolution is true. . ." In the fall of 1981, Patterson addressed the Systematics Discussion Group at the American Museum of Natural History. Once a month, the group meets in an upstairs classroom at the museum, opposite the dinosaur exhibit hall. The audience in any given month is likely to be made up of museum staff, graduate students from nearby universities, and the occasional amateur like Norman Macbeth, the author of *Darwin Retried.* (Systematics is a science of classification; taxonomists working in systematics study the way taxonomic groups relate to one another in nature.) There may be no more than fifteen people on hand when the discussion focuses on, say, fossil rodent teeth; or there may be 150 or more when Richard C. Lewontin, the renowned geneticist and author, gives a talk on the meaning (if any) of adaptation in biology

4 Patterson's address was titled "Evolutionism and Creationism." Patterson is not a creationist, but he had been trying to think like one as a sort of experiment. "It's true," he told his audience, "that for the last eighteen months or so I've been kicking around non-evolutionary or even anti-evolutionary ideas." He went on:

> I think always before in my life when I've got up to speak on a subject I've been confident of one thing—that I know more about it than anybody in the room, because I've worked on it. Well, this time it isn't true. I'm speaking on two subjects, evolutionism and creationism, and I believe it's true to say that I know nothing whatever about either of them.
>
> One of the reasons I started taking this anti-evolutionary view, or let's call it a non-evolutionary view, was that last year I had a sudden realization. For over twenty years I had thought I was working on evolution in some way. One morning I woke up and something had happened in the night, and it struck me that I had been working on this stuff for more than twenty years, and there was not one thing I knew about it. It's quite a shock to learn that one can be so misled for so long. Either there was something wrong with me or there was some-

> thing wrong with evolutionary theory. Naturally I know there is nothing wrong with me, so for the last few weeks I've tried putting a simple question to various people and groups.
>
> Question is: Can you tell me anything you know about evolution? Any one thing, any one thing that is true?

5 In the public mind, challenges to Darwin's theory of evolution are associated with biblical creationists who periodically remove their children from schoolrooms where they are being taught that man evolved from monkeys. Most Americans know about the Scopes trial of 1925, in which a Tennessee high school teacher was fined $100 for teaching evolutionary theory. Four years ago there was the trial in San Diego in which Kelly Seagraves, director of the Creation Science Research Center, unsuccessfully sued the state of California over regulations governing the teaching of evolution in California public schools. (Seagraves wanted science teachers to be required to mention pertinent passages from the Book of Genesis.) What most people do not know is that for much of this century, and especially in recent years, scientists have been fighting among themselves about Darwin and his ideas.

6 Scientists are largely responsible for keeping the public in the dark about these in-house arguments. When they see themselves as beleaguered by opponents outside the citadel of science, they tend to put their differences aside and unite to defeat the heathen. The layman sees only the closed ranks. At the moment, with creationism apparently quiescent, we can, if we listen hard enough, hear fresh murmurs of dissent within the scientific walls. These debates are more complicated, perhaps, than the old contest, Science vs. Religion, but they are at least as interesting, and sometimes as heated.

7 One of the least publicized and least understood challenges to Darwin and the theory of evolution—and surely one of the more fascinating, in its sweep and rigor—involves a school of taxonomists called cladists. (A "clade" is a branch, from the Greek *klados*; "cladist" is pronounced with a long *a*.) Particularly interesting—vexing, evolutionary biologists would say (and do)—are those who toil in what is called transformed cladistics, and who might be thought of as agnostic evolutionists. Like many who have broken with a faith and challenged an orthodoxy, the transformed cladists are perhaps best defined by an opponent—in this case, the British biologist Beverly Halstead. Asked

not long ago in a BBC interview what he thought of transformed cladistics, Halstead replied: "Well, I object to it! I mean, this is going back to Aristotle. It is not pre-Darwinian, it is Aristotelian. From Darwin's day to the present we've understood there's a time element; we've begun to understand evolution. What they are doing in transformed cladistics is to say, Let's forget about evolution, let's forget about process, let's simply consider pattern."

8 Since Darwin's time, biologists have been absorbed in process: Where did we come from? How did everything in nature get to be what it now is? How will things continue to alter? The transformed cladists—they are sometimes called pattern cladists—are not concerned primarily with time or process. To understand why, it helps to know that they are trained in taxonomy: they are rigorous, scrupulous labelers. Their job as taxonomists is to discover and name the various groups found in nature—a task first assigned to Adam by God, according to Genesis—and put them into one category or another. Taxonomists try to determine not how groups came into existence but what groups exist, among both present-day and fossil organisms. To understand that cladists believe this knowledge must be acquired before ideas about process can be tested is to understand the natural tension that exists between taxonomists and evolutionary biologists.

9 The transformed cladists have escalated the battle. In the 1940s and 1950s, years which witnessed the growth of evolutionary biology, taxonomists allowed themselves what might be called a bit of artistic license. (They called it the new systematics.) This occurred in part, no doubt, because taxonomy had come to be thought of as dull and stuffy—particularly by evolutionists like Sir Julian Huxley (the grandson of Darwin's contemporary champion Thomas Henry Huxley), who believed it was high time to cease being "bogged down in semantics and definitions." (Sir Julian said this in 1959 at the University of Chicago during a centennial celebration for Darwin's *On the Origin of Species.*) Taxonomists, in other words, were regarded as bookkeepers and accountants in need of a little loosening up. In his 1959 book *Nature & Man's Fate*, Garrett Hardin, a professor of human ecology at the University of California at Santa Barbara, quoted a zoologist as giving this advice: "Whoever wants to hold to firm rules should give up taxonomic work. Nature is much too disorderly for such a man."

10 The transformed cladists think otherwise, and have sought to

reestablish taxonomic rigor. In doing so, they have come to think that it is the evolutionists who have the problem—the problem being slipshod methodology. Colin Patterson, perhaps the leading transformed cladist, has enunciated what might be regarded as the cladists' battle cry: "The concept of ancestry is not accessible by the tools we have." Patterson and his fellow cladists argue that a common ancestor can only be hypothesized, not identified in the fossil record. A group of people can be brought together for a family reunion on the basis of birth documents, tombstone inscriptions, and parish records—evidence of process, one might say. But in nature there are no parish records; there are only fossils. And a fossil, Patterson told me once, is a "mess on a rock." Time, change, process, evolution—none of this, the cladists argue, can be read from rocks.

11 What can be discerned in nature, according to the cladists, are patterns—relationships between things, not between eras. There can be no absolute tracing back. There can be no certainty about parent-offspring links. There are only inferences drawn from fossils. To the cladists, the science of evolution is in large part a matter of faith—faith different, but not all *that* different, from that of the creationists.

12 "I really put my foot in it," Patterson told me that day I first met him nearly two years ago. We were in a restaurant on Columbus Avenue near the Museum of Natural History, and he was recalling the talk he had given eighteen months earlier to the systematics discussion group. "I compared evolution and creation and made a case that the two were equivalent. I was all fired up, and I said what I thought. I went through merry hell for about a year. Almost everybody except the people at the museum objected. Lots of academics wrote. Deluges of mail. 'Here we are trying to combat a political argument,' they said, 'and you give them ammunition!'"

13 He ordered something from the menu and said: "One has to live with one's colleagues. They hold the theory very dear. I found out that what you say will be taken in 'political' rather than rational terms."

14 Patterson told me that he regarded the theory of evolution as "often unnecessary" in biology. "In fact," he said, "they could do perfectly well without it." Nevertheless, he said, it was presented in textbooks as though it were "the unified field theory of biology," holding the whole subject together—and binding the profession to it. "Once something has that status," he said, "it becomes like religion."

15 The founding father of cladistics was an entomologist named Willi Hennig. Hennig was born in what is now East Germany and spent the bulk of his career there, studying and classifying flies. At some point in the mid-1960s (there is very little biographical information available about him) he turned up in West Germany; he died there, at the age of 63, in 1976. His principal work is *Phylogenetic Systematics*, an updated version of which was translated into English and published in the United States in 1966 by the University of Illinois Press. It is a difficult book, and an enormously influential one. By the 1970s, as the prominent evolutionary biologist Ernst Mayr wrote in *The Growth of Biological Thought,* a virtual Hennig cult had developed. A Willi Hennig Society was formed in 1980, and its fourth annual meeting, held last summer in London, was attended by some 250 scientists from around the world. Last month, the society published the first issue of its new quarterly journal, *Cladistics.* According to David Hull, the philosopher of science (he was at the meeting too), "among evolutionary biologists, cladistics is what everyone is arguing about."

16 At the heart of cladistics are the concepts of "plesiomorphy" and "paraphyly." A characteristic, or trait, is said to be plesiomorphic if it is found in a group of organisms of more general scope than the specific group under consideration. Thus, while all primates have hair, hair is also a characteristic of a more general class of creatures—mammals. What Hennig called the fallacy of plesiomorphy is the belief that a characteristic (like hair) identifies and helps to define a particular species or order of animal life when in fact it can be found among a broader group.

17 Hennig also objected to the still common practice in biology of identifying a grouping of animal life only by the absence of certain characteristics. (His reasoning was Aristotelian; in *On the Parts of Animals,* Aristotle wrote that "there can be no specific forms of a negation, of Featherless, for instance or of Footless, as there are of Feathered or Footed.") It was the lack of precision that bothered Hennig: a feathered animal is one thing (a bird); a non-feathered animal is anything (except a bird). Groups in nature defined by an absence of characteristics Hennig called paraphyletic.

18 By calling attention to the paraphyletic traits, Hennig helped revive the rigor taxonomy once prided itself on. Colin Patterson and other transformed cladists have moved on to examine—and call into

question—the crucial role that paraphyletic groups and species play in evolutionary theory. In his 1981 talk at the Museum of Natural History, for example, Patterson touched on the subject of invertebrates. Invertebrates make up one of the two general categories of animals. The grouping comprises a huge and often bewildering diversity of animals, from the simplest single-cell protozoan to insects, clams, worms, and crabs. Every schoolchild learns that what brings this wide array of creatures under one heading is their shared lack of a backbone. Cladists like Patterson have asked: *Why* group them this way? What function does it serve? The problem they have is this: the term *invertebrate* does not serve a scientific function; it is too nebulous, too inexact for that. (It also accurately describes strawberries and chairs.) What the term *invertebrate* does serve, the cladists maintain, is a rhetorical function: it makes possible the claim, found in many textbooks, that "vertebrates evolved from invertebrates." According to the cladistic reading, the last two words of the four-word statement do not contain any information that is not asserted as factual by the first two words; "vertebrates evolved" simply means that the first vertebrate had parents without backbones. The transformed cladists claim that "vertebrates evolved from invertebrates" is a disguised tautology.

19 In his museum talk, Patterson said that groups defined only by negative traits have "no existence in nature, and they cannot possibly convey knowledge, though they appear to when you first hear them." Evolutionary biologists maintain that negatively defined groups make sense and serve a purpose; they tend to accuse the cladists, as one writer recently did in the magazine *Science*, of engaging in "verbal legerdemain." But Patterson and his colleagues point their fingers back at the evolutionists. Patterson for one has called the paraphyletic groups "voids."

20 What evolutionary theory does, the cladists say, is make claims about something that cannot be determined by studying fossils. They say that the "tree of life," with its paraphyletic branches, is nothing more than a hypothesis, a reasonable guess.

21 Nor do they believe it will ever be anything more than that. When asked about this in an interview, Patterson said: "I don't think we shall ever have any access to any form of tree which we can call factual." He was then asked: "Do you believe it to be, then, no reality?" He replied: "Well, isn't it strange that this is what it comes to, that you have to ask me whether I believe it, as if it mattered whether I believe

it or not. Yes, I do believe it. But in saying that, it is obvious it is faith."

22 Cladists do not spend their time on the lecture circuit drumming up sentiment against Darwin. Some of them would like it if all the talk about evolution just quietly went away. Evolution is not important to the work they do. That work involves finding the positive and verifiable characteristics of the various species and determining how all these species fit together in the animal kingdom—what patterns exist in nature. Their interest is the here and now, not how it all came to be.

23 I recently spent some time with two cladists on the staff of the Museum of Natural History. I first met with Gareth Nelson, who in 1982 was named chairman of the department of ichthyology. Nelson graduated from the University of Hawaii in 1966 and he joined the museum staff a year later. The walls of Nelson's office were lined with boxes of articles from scientific journals, and a large table was covered with papers and jars stuffed with small, silvery fish preserved in alcohol: anchovies. Nelson is just about the world's expert on anchovies, although he told me that the number of people studying them (three or four) is much smaller than the number of anchovy species (there are 150 known species, and Nelson believes there are many more). This disparity between the magnitude of the scientific "problem" and the number of people working on it is a commonplace in biology. Most laymen think that the experts have pretty exhaustively studied the earth's biota, when they have barely scratched the surface.

24 Nelson put the issue of evolution this way: in order to understand what we actually know, we must first look at what it is that the evolutionists claim to know for certain. He said that if you turn to a widely used college text like Alfred Romer's *Vertebrate Paleontology*, published by the University of Chicago Press in 1966 and now in its third edition, you will find such statements as "mammals evolved from reptiles," and "birds are descended from reptiles." (Very rarely, at least in the current literature, will you find the claim that a given species evolved from another given species.) The trouble with general statements like "mammals evolved from reptiles," Nelson said, is that the "ancestral groups are taxonomic artifacts." These groups "do not have any characters that are unique," he said. "They do not have defining characters, and therefore they are not

real groups." I asked Nelson to name some of these allegedly "unreal" groups. He replied: invertebrates, fishes, reptiles, apes. According to Nelson, this does not by any means exhaust the list of negatively defined groups. Statements imputing ancestry to such groups have no real meaning, he said.

25 I asked Nelson about the fossil record. Don't we know that evolutionary theory is true from the fossils? Like most people, I thought the natural history museums had pretty well worked out the fossil sequences, much as in an automobile museum you can find the "ancestors" of contemporary cars lined up in sequence: Thunderbird back to Model T.

26 "Usually with fossils all you find are a few nuts and bolts," Nelson said. "An odd piston ring, maybe, or different pieces of a carburetor that are spread out or piled on top of one another, but not in their correct arrangement."

27 He maintained that too much importance has been attached to fossils. "And it's easy to understand why," he said. "You put in all this effort studying them, and you get out a little bit. Therefore you are persuaded that that little bit must be very important. I can get ten times more information per unit with recent fishes. So if you put in all that effort on fossils, you are inclined to say that the information you get is worth ten times as much."

28 Nelson said it was quite common for paleontologists to go to all the trouble of digging up fossils without realizing that the animals in question were still walking about. (Think of spending months hunting for a book in used-book stores without realizing it was still in print.) "Say you dig up a 50-million-year-old beetle," he said. "It looks like it belongs to a certain family, but there may be 30,000 species in the family. What do you do? Go through all 30,000? No, you just give it an appropriate-sounding name, *Eocoleoptera,* say. If it is a species that has been in existence for 50 million years, somebody else will have to find that out, because you don't have enough time. You're out digging in the rocks, not poking through beetle collections in museums."

29 I asked him about anchovy fossils. How far back do they go? "Well," he said, "Lance Grande, who was a student here recently, studied that, and it turns out that all the fossils previously described as anchovies are not anchovies at all." (Grande is now an assistant curator in the department of geology at the Field Museum of Natural History

in Chicago.) "In other words," Nelson said, "the people who described them did not do a very good job. So the fossil record of anchovies was reduced to zero. However, there was something in the British Museum that I think Colin Patterson told Grande about, something from the Miocene in Cyprus; maybe 10 million years old. And it turned out to be an anchovy—the only known fossil. It has not yet been described in detail, but there is information suggesting it is the same kind of animal we find inhabiting the Mediterranean today."

30 A week or two after I met with Nelson I spoke to Norman Platnick, a curator in the museum's entomology department and an expert on spiders. On my way to see him on the fifth floor, I was joined in the elevator by a couple of lab assistants who were wheeling on a cart what looked like a dinosaur head. (I was reminded that for a long time the museum had the wrong head on its brontosaurus. One of the few bits of conventional wisdom about paleontology is that entire animals can be reconstructed from scraps of bone. Paleontologists now repudiate the idea, first enunciated by the French anatomist Baron Cuvier in the early 1800s. Steve Farris, a professor in the department of ecology and evolution at the State University of New York at Stony Brook and the president of the Hennig Society, told me that Cuvier erected a monument to his own error in the form of a cement statue of an iguanodon, now at the Crystal Palace outside London. "The animal that Cuvier imagined was four-footed and resembled a rhinoceros," Farris said. "The complete skeleton of the iguanodon is now known—the animal was bipedal, with a long tail." As for the idea that the relationship of early animals to present-day ones is well established, Farris said: "When they are writing for a general audience, a lot of paleontologists do try to give that impression.")

31 Not far from the elevator I found Platnick's orderly office: spiders (dead) inside little labeled bottles; book-filled shelves; journal articles neatly stacked. It would seem that professional biologists spend at least as much time studying each other's work as they do the world around them.

32 Platnick, who is rather square-shaped and bearded, told me that when he was an undergraduate at a small Appalachian college, he would go along with his wife when she collected millipedes. "But I was a wretched millipede collector," he said. "When we arrived home, all I would have in my jars would be spiders." So he started to study them. Today he has a Ph.D. from Harvard, and he and Nelson are co-

authors of a book recently published by Columbia University Press entitled *Systematics and Biogeography: Cladistics and Vicariance.*

33 Spiders, which go back to the Devonian period, 400 million years ago, belong to the class Arachnida and the phylum Arthropoda. They are among the "invertebrates," in other words, and are not well preserved in the fossil record. About 35,000 species of spiders have been identified, Platnick said, "but there may be three times that many in the world." He thought there were perhaps four full-time systematists examining spiders in the United States, "and perhaps another dozen who teach at small colleges and do some research." There is an American Arachnological Society, with 475 members worldwide, some of them amateurs. They meet once a year and discuss scorpions and daddy-long-legs, as well as spiders.

34 "Most of the spiders I look at may have been looked at by two or three people in history," Platnick said, adding that he would most likely be dead before anyone looked at them again.

35 I asked Platnick what was known about spider phylogeny, or ancestry.

36 "Very little," he said. "We still don't know a hill of beans about that." We certainly don't know, he said, what species the animal belonged to that was the ancestor of the very first spider. All we know of such an animal is that it was *not* a spider. We don't even know of any links in the (presumed) 400-million-year chain of spider ancestry.

37 "I do not *ever* say that this spider is ancestral to that one," Platnick said firmly.

38 "Does anyone?"

39 "I don't know of a single case in the modern literature where it's claimed that one spider is the ancestor of another."

40 Some spiders have been well preserved in amber. Even so, Platnick said, "very few spider fossils have been so well preserved that you can put a species name on them." After a pause he added: "You don't learn much from fossils."

41 In view of Platnick's comments about our knowledge of spider ancestry, I was curious to know what he thought of the following passage from a well-known high school biology text, *Life: An Introduction to Biology*, by George Simpson and William S. Beck, first published in 1957 by Harcourt Brace Jovanovich and still in print.

> An animal is not classified as an arachnid because it has four or five pairs of legs rather than three. It is classified in the Arachnida because it has the same ancestry as other arachnids, and a different ancestry from insects over some hundreds of millions of years, as attested by all the varying characteristics of the two groups and by large numbers of fossil representatives of both.

42 At that he threw himself back in his chair, and burst out laughing.

43 In this passage, Simpson and Beck were practicing the verbal sleight of hand that has been common in evolutionary biology since the 1940s. All we know for sure is that there is a group of organisms (in this case spiders) that are identifiable as a group because they have certain unique characteristics. They have spinnerets for spinning silk, for instance, and thus we can say that all organisms with spinnerets are spiders. (They share other unique features, too.)

44 If we want to explain *why* thousands of members of a group have features uniquely in common, that is another matter entirely. We can, if we like, posit a theoretical common ancestor in the ur-spider, which transmitted spider traits to all its descendants. That is precisely what Darwin did in *On the Origin of Species.* But Simpson and Beck do something very different. They say that the composition of the class Arachnida was determined by examining not the features of spiders but their *ancestral lines.* But no such pedigrees are known to science—not just with respect to spiders but with respect to *all* groups of organisms.

45 The point stressed by the cladists is this: unless we know the taxonomic relationships of organisms—what makes each unique and different from the other—we cannot possibly guess at the ancestral relationships. Things in nature here and now must be ranked according to their taxonomic relationship before they can be placed in a family tree. Thus the speculations of evolutionists ("Do X and Y have a common ancestor?") must be subordinate to the findings of taxonomists ("X and Y have features not shared by anything else"). If fossils came with pedigrees attached, this laborious method of comparison would not be necessary; but of course they don't.

46 "Stephen Jay Gould does his work without bothering about cladistics, I assure you," Platnick said, citing a recent paper by Niels Bonde, a paleontologist at the University of Copenhagen. Platnick went on to say that "the literature is replete with such statements as

'fossil X is the ancestor of some other taxon,' when it has not even been shown that fossil X is the closest relative of that taxon." (By "closest relative" he means that the two taxa form a group having unique characteristics.) "This is seen most commonly in accounts of human paleontology, but it is by no means restricted to it," Platnick said.

47 One reason why many laymen readily accept evolution as fact is that they have seen the famous "horse sequence" reproduced in textbooks. The sequence, which shows a gradual increase in the size of the horse with time, is dear to the hearts of textbook writers, in large part because it is on display at the American Museum of Natural History. For obvious reasons, the museum staff are uncomfortable going on record about the horse sequence, but when Niles Eldredge, a curator in the department of invertebrates at the museum and co-author, with Stephen Jay Gould, of the "punctuated equilibria" theory of evolution (organisms stay the same for millions of years, then change quickly rather than gradually, as Darwin believed), was asked about it once, he said:

> There have been an awful lot of stories, some more imaginative than others, about what the nature of that history [of life] really is. The most famous example, still on exhibit downstairs, is the exhibit on horse evolution prepared perhaps fifty years ago. That has been presented as the literal truth in textbook after textbook. Now I think that that is lamentable, particularly when the people who propose these kinds of stories may themselves be aware of the speculative nature of some of that stuff.

48 When I brought the subject up with Platnick, he said that he thought horse fossils had not yet been properly classified, or even exhaustively studied. I wanted to know whether Platnick believed that evolution has occurred. He said he did, and that the evidence was to be found in the existing hierarchical structure of nature. All organisms can, as it were, be placed within an internested set of "boxes." The box labeled "gazelles" fits in the larger box labeled "ungulates" (animals with hoofs), which fits inside the "mammals" box, which fits inside "tetrapods" (four-footed animals), which fits inside "vertebrates." The grand task of taxonomy, Platnick said, is to describe this hierarchical pattern precisely, and in particular to define the traits that delineate the boundaries of each "box."

49 Whether taxonomy will ever fill in all the blanks in the pattern

is a question Platnick cannot answer. One problem, he said, is the shortage of taxonomists. "Systematics," he said, "doesn't have the glamour to attract research funds." Research grants have increasingly gone to molecular and biochemical studies; the result is that support for taxonomy at many institutions has, he said, "withered away." This bothered Platnick. "I am fully prepared to stand up to any biologist who says evolutionary theory is more important, or more basic. Without the results of systematics there is nothing to be explained."

50 I wanted to find out what those on the other side—the evolutionary biologists and paleontologists—had to say about what the cladists are saying. First I went to the bookshelf. In his 1969 book *The Triumph of the Darwinian Method* (recently reprinted by the University of Chicago Press), Michael T. Ghiselin, one of Darwin's greatest admirers, seems to be taking on the cladists (or trying to) when he writes:

> Instead of finding patterns in nature and deciding that because of their conspicuousness they seem important, we discover the underlying mechanisms that impose order on natural phenomena, whether we see that order or not, and then derive the structure of our classification system from this understanding.

51 I next looked in *Hen's Teeth and Horse's Toes*, Stephen Jay Gould's volume of essays on natural history. "No debate in evolutionary biology has been more intense during the past decade than the challenges raised by cladistics against traditional schemes of classification," Gould writes. He is not sympathetic to cladistics ("its leading exponents in America are among the most contentious scientists I have ever encountered"), but in his essay "What, If Anything, Is a Zebra?" he admits that "behind the names and the nastiness lies an important set of principles." These he enunciates, only to repudiate. He acknowledges that a strict taxonomy would eliminate groups like apes and fishes. But when cladists go this far, "many biologists rebel, and rightly, I think." Like his Harvard colleague Edward O. Wilson, the Frank B. Baird Professor of Science, Gould opts for the "admittedly vague and qualitative, but not therefore unimportant notion of overall similarity" of form.

52 I decided it would be a good idea to talk with a scientist who believes strongly in evolutionary theory. Last May, I traveled to Boston

to meet with Richard C. Lewontin, a geneticist, a one-time president of the Society for the Study of Evolution, a well-known writer on science, and currently Agassiz Professor of Zoology at Harvard. I had seen a quote from Lewontin used as a chapter head in a book titled *Science on Trial*, by Douglas Futuyma. The quote, as edited, read: "Evolution is fact, not theory. . . . Birds evolve from nonbirds, humans evolve from nonhumans."

53 Lewontin was uncharacteristically attired in a scientist's regulation white lab coat when I first saw him (instead of his usual blue work shirt). We talked a bit about his stand against biological determinism. Finally it was time to get around to the point of my visit. What about these claims: evolution is fact; birds evolve from nonbirds, humans from nonhumans? The cladists disapproved, I said.

54 He paused for a split second and said: "Those are very weak statements, I agree." Then he made one of the clearest statements about evolution I have heard. He said; "Those statements flow simply from the assertion that all organisms have parents. It is an empirical claim, I think, that all living organisms have living organisms as parents. The second empirical claim is that there was a time on earth when there were no mammals. Now, if you allow me those two claims as empirical, then the claim that mammals arose from non-mammals is simply a conclusion. It's the deduction from two empirical claims. But that's all I want to claim for it. You can't make the direct empirical statement that mammals arose from non-mammal."

55 Lewontin had made what seemed to me to be a deduction—a materialist's deduction. "The only problem is that it appears to be based on evidence derived from fossils," I said. "But the cladists say they don't really have that kind of information."

56 "Of course they don't," Lewontin said. "In fact, the stuff I've written on creationism, which isn't much, has always made that point. There is a vast weight of empirical evidence about the universe which says that unless you invoke supernatural causes, the birds could not have arisen from muck by any natural processes. Well, if the birds couldn't have arisen from muck by any natural processes, then they had to arise from non-birds. The only alternative is to say that they did arise from muck—because God's finger went out and touched that muck. That is to say, there was a non-natural process. And that's really where the action is. Either you think that complex organisms arose by non-natural phenomena, or you think that they arose by natural phenomena. If they

arose by natural phenomena, they had to evolve. And that's all there is to it. And that's the only claim I'm making."

57 He reached for a copy of his 1982 book *Human Diversity*, and said: "Look, I'm a person who says in this book that we don't know anything about the ancestors of the human species." (He writes on page 163: "Despite the excited and optimistic claims that have been made by some paleontologists, no fossil hominid species can be established as our direct ancestor. . . .") "All the fossils which have been dug up and are claimed to be ancestors—we haven't the faintest idea whether they are ancestors. Because all you've got, and the cladists are right. . ." He got up and began to do his famous rat-a-tat-tat with a piece of chalk on the blackboard. "All you've got is Homo Sapiens there, you've got *that* fossil there, you've got another fossil there. . . this is time here. . . and it's up to you to draw the lines. Because there *are* no lines. I don't think any one of them is likely to be the direct ancestor of the human species. But how would you know it's *that* [pat] one?

58 "The only way you can know that some fossil is the direct ancestor is that it's so human that it *is* human. There is a contradiction there. If it is different enough from humans to be interesting, then you don't know whether it's an ancestor or not. And if it's similar enough to be human, then it's not interesting."

59 He returned to his chair and looked out at the slanting rain. "So," he said. "Look, we're not ever going to know what the direct ancestor is."

60 What struck me about Lewontin's argument was how much it depended on his premise that all organisms have parents. In a sense, his argument includes the assertion that evolutionary theory is true. Lewontin maintains that his premise is "empirical," but this is so only in the (admittedly important) sense that it has never to our knowledge been falsified. No one has ever found an organism that is known not to have parents, or a parent. This is the strongest evidence on behalf of evolution.

61 Our belief, or "faith," that, as Patterson says, "all organisms have parents" ultimately derives from our acceptance of the philosophy of materialism. It is hard for us to understand (so long has materialism been the natural habitat of Western thought) that this philosophy was not always accepted. In one of his essays on natural history reprinted in *Ever Since Darwin*, Stephen Jay Gould suggests that Darwin delayed publishing his theory of evolution by natural selection because he was,

perhaps unconsciously, waiting for the climate of materialism to become more firmly established. In his 1838 *M Notebook* Darwin wrote: "To avoid stating how far, I believe, in Materialism, say only that emotions, instincts, degrees of talent, which are hereditary are so because brain of child resembles parent stock." Darwin realized that the climate *had* changed—that evolution was "in the air"—in 1858 when he was jolted by Alfred Russel Wallace's paper outlining a theory of the mechanism of evolution very similar to his own.

62 The theory of evolution has never been falsified. On the other hand, it is also surely true that the positive evidence for evolution is very much weaker than most laymen imagine, and than many scientists want us to imagine. Perhaps, as Patterson says, that positive evidence is missing entirely. The human mind, alas, seems on the whole to find such uncertainty intolerable. Most people want certainty in one form (Darwin) or another (the Bible). Only evolutionary agnostics like Patterson and Nelson and the other cladists seem willing to live with doubt. And that, surely, is the only truly scientific outlook.

Suggestions for Discussion and Writing

1. What stereotypes do you have of people who oppose Darwin's theories of evolution? How does Bethell play off those stereotypes in this essay?

2. What happens when people define something by what is *not* there? Can you think of any concepts (not necessarily those in natural history) that our society habitually defines by what's missing? What are the advantages and disadvantages of such definitions?

3. What do you think Bethell's purpose is in writing this essay? Do you think Bethell believes in evolution? What evidence can you find in the essay to support your answer?

4. Why do you think scientists like to keep their disagreements "in the family"? Do you think such disputes are limited to science? What happens to a community when its most valued beliefs are challenged?

5. Imagine you have been asked to moderate a debate about evolutionary theory between Eldredge and Bethell. What questions would you ask? On what points would you expect them to differ? Cast your answer as your introductory remarks to whatever audience you think would attend such a debate.

Bart J. Bok

A Critical Look at Astrology

BART J. BOK was born in the Netherlands in 1906, but moved to the United States soon after. From 1929-1957, he was a professor of astronomy at Harvard. Then, after ten years as director of the Australian National University's Mt. Stromlo Observatory, he moved to the University of Arizona, where he taught until his retirement in 1974. With his wife Priscilla Fairfield Bok, he was the author of *The Milky Way*. Its fifth edition was published in 1981, forty years after its original appearance; it remains the definitive source for information on our galaxy. He is also the author of *Objections to Astrology* (1975), which he wrote with Lawrence E. Jerome.

Bok is noted for making complex subjects understandable to average readers. He told one reviewer how he suited material to his audiences. "Priscilla and I wanted to write a book aimed specially at beginning university students and at bright boys and girls of high school age. Our grandchildren were fine test objects when judging what to include and how to approach each subject." Bok died in 1984. This essay originally appeared in *The Humanist* in 1975.

During the past ten years, we have witnessed an alarming increase in the spread of astrology. This pseudoscience seems to hold fascination especially for people of college age who are looking for firm guideposts in the confused world of the present. It is not surprising that people believe in astrology when most of our daily newspapers regularly carry columns about it and when some of our universities and junior colleges actually offer astrology courses. The public, young and old, has the right to expect from its scientists, especially from astronomers, clear and clarifying statements showing that astrology lacks a firm scientific foundation.

2 I have spoken out publicly against astrology every ten years or so, beginning in 1941 in "Scientists Look at Astrology," written with Margaret W. Mayall for the now-defunct *Scientific Monthly* (Volume 52). I have softened a bit since my early crusading days, for I have come to realize that astrology cannot be stopped by simple scientific argument only. To some it seems almost a religion. All I can do is state clearly and unequivocally that modern concepts of astronomy and space physics give no support—better said, negative support—to the tenets of astrology.

3 Not more than a dozen or so of my fellow astronomers have spoken out publicly on astrology. Twice I suggested to my friends on the Council of the American Astronomical Society that the council issue a statement pointing out that there is no scientific foundation for astrological beliefs. Both times I was turned down, the principal argument being that it is below the dignity of a professional society to recognize that astrological beliefs are prevalent today. To me it seems socially and morally inexcusable for the society not to have taken a firm stand. Astronomers as a group have obviously not provided the guidance that the public sorely needs. Those who live in a free society are entitled to believe in whatever causes they care to espouse. However, I have had more than half a century of day-to-day and night-to-night contacts with the starry heavens, and it is my duty to speak up and to state clearly that I see no evidence that the stars and planets influence or control our personal lives and that I have found much evidence to the contrary.

The Origins of Astrology

4 Astrology had its origins in the centuries before the birth of Christ. Present-day astrological concepts and techniques largely go back to the period 100 to 200 A.D. It was only natural that early civilizations would consider the stars and planets in the heavens as awesome evidence of supernatural powers that could magically affect their lives. Variety was brought into the picture by the constantly changing aspects of the heavens. No one can blame the Egyptians, the Greeks, the Arabs, or the people of India for having established systems

of astrology at times when they were also laying the foundations for astronomy. Right up to the days of Copernicus, Galileo, and Kepler (who was an expert astrologer)—even to the time of Isaac Newton—there were good reasons for exploring astrology.

5 However, all this changed when the first measurements were made of the distances to the sun, planets, and stars and when the masses of these objects were determined. The foundations of astrology began to crumble when we came to realize how vanishingly small are the forces exerted by the celestial objects on things and people on earth—and how very small are the amounts of radiation associated with them received on earth. The only perceptible and observable effects evident to all of us are produced by the tidal forces caused by the gravity of the moon and sun. To assume that the sun, moon, and planets would exert special critical forces upon a baby at birth—forces that would control the future life of the infant—seems to run counter to common sense. Radiative effects are also dubious. It is even less likely that the stars—each one a sun in its own right and several hundred thousand or more times farther from the earth than our sun—would exercise critical effects on a baby at birth. Some seasonal effects there might well be, for a baby born in northern latitudes in April faces initially a warm summer period; one born in October, a cool winter season.

6 Before the days of modern astronomy, it made sense to look into possible justifications for astrological beliefs, but it is silly to do so now that we have a fair picture of man's place in the universe.

Horoscopes: Their Preparation and Interpretation

7 Astrology claims to foretell the future by studying the positions of the sun, moon, and planets in relation to the constellations of stars along the celestial zodiac at the time of the birth of the subject. This is done through the medium of the *horoscope.* Anyone with a knowledge of beginning astronomy and with an American Nautical Almanac on his desk can proceed to draw one. I have found it a not unpleasant pastime on several occasions. Some of our readers may be interested in learning a little about the technical procedures involved in the preparation of a horoscope. Figures 1 and 2, which are reproduced from the previously mentioned *Scientific Monthly* article, may help to illustrate the procedures.

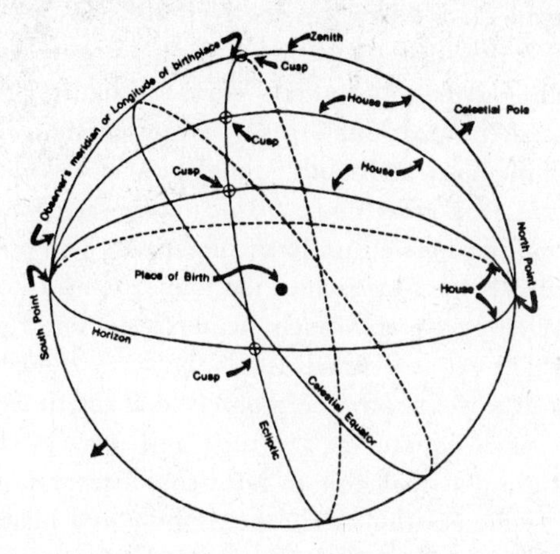

Figure 1. *The Celestial Sphere Divided into Twelve Parts*

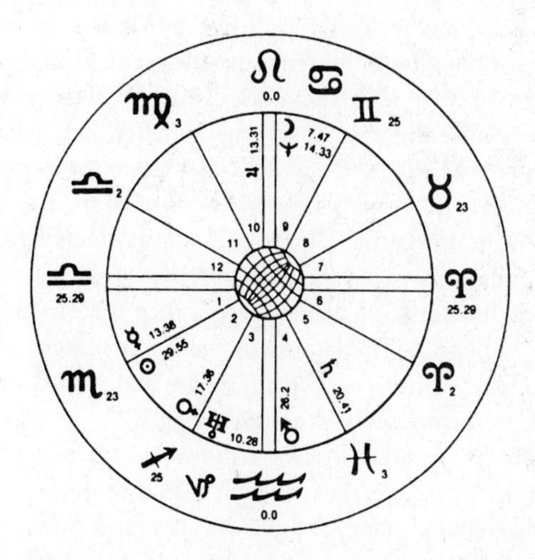

Figure 2. *A Conventional Type of Natal Horoscope. The horoscope is drawn for November 23, 1907, 4 a.m. Eastern-Standard Time, latitude 40°43'N, longitude 73°58'W. The spokes of the wheel mark the limits of the houses; the zodiacal signs and the degrees mark the cusps. The position of the sun, moon, and planets are shown by their symbols.*

8 For a given place on earth—the birthplace of the subject—the celestial sphere is drawn in the standard manner favored by teachers of college beginning-astronomy courses. We see in figure 1 that the local celestial meridian (a great circle passing through the celestial poles, the zenith, and the north and south points on the horizon) and the local celestial horizon (the great circle 90 degrees away from the zenith overhead) divide the sphere into four equal parts. The celestial sphere is further divided by cutting each of the four sections into three equal slices by great circles passing through the north and south points on the local horizon. The ecliptic, which traces the sun's annual path across the sky, is just as it would have been observed at the time and place of birth. The celestial equator at the time and place of birth is also shown, but it plays a small role in horoscope preparation, except to help in plotting the position of the sun, moon, and planets from the Almanac. The intersections between the twelve great circles and the ecliptic circle mark the *cusps* of the twelve *houses.*

9 We are now ready to draw the horoscope for the subject at the time and place of birth. We see in figure 2 that the houses and their cusps are drawn on a graph representing the plane of the ecliptic. This is accomplished either with the aid of suitable astrological tables or by the use of some simple spherical trigonometry. The whole business can, if so desired, be prepared nicely as a program for a reasonably fast computer, all of which helps create the impression that astrology is basically scientific in nature. The houses are numbered from 1 to 12 (as shown in figure 2), with house 1 being the one that is about to rise above the local horizon. Standard tables are then used to mark the positions of the zodiacal constellations in the outer margin of the horoscope wheel. The positions of the sun, moon, and planets are shown by their symbols in the houses where they belong.

10 To sum up: Figure 2 shows a horoscope in which the twelve houses, with their cusps marked, and the positions of the zodiacal constellations and the sun, moon, and planets are drawn on the plane of the ecliptic, just as they would have been observed at the time and place of birth of the subject.

11 The type of horoscope shown in figure 2 is a *natal* horoscope. It becomes—according to astrologers—the all-important guide to predicting a person's future. There are other types of horoscopes in use—*judicial* and *hororary* ones, for example—but these need not concern us here.

12 The abracadabra begins when the astrologer starts to interpret a person's horoscope. The first and most important item is the date of birth, which makes a person an Aries if born between March 21 and April 19, a Taurus if born in the month following, and so on. The date of birth naturally fixes the sign of the zodiac in which the sun is located. It is the important fact that tells whether a person is an Aries, a Pisces, or whatever.

13 The next important item to note is if the moon or the sun and what planets are in the first house, the one that is about to rise above the horizon at the time and place of birth. The *ascendant* is defined as the sign of the zodiac that is associated with the first house. It is obvious that the time of birth must be precisely known if a proper horoscope is to be prepared.

14 Third in line are the positions relative to each other of the sun, moon, and planets in the various houses. The so-called *aspects* are noted. These indicate which celestial objects are in conjunction (near to each other in position), and which are 60, 90, 120, or 180 degrees apart in the heavens. These aspects are important for astrological interpretation, as are the positions of the planets in the houses in which they are found.

15 The astrologer can refine his interpretations to any desired extent—the end product becoming increasingly more expensive as further items are added.

16 How and by whom were these rules of analysis and interpretation of horoscopes first established? They go back to antiquity, basically to the work of Ptolemy in the second century. Ptolemy wrote two famous books: *Almagest,* the most complete volume on the motions of the planets published during the great days of Greek astronomy, and *Tetrabiblos,* the bible of astrology. The *Almagest* is today treated with respect and admiration by historians of science, and it is clearly one of the great works of the past. However, no astronomer would think of referring to it today when considering problems relating to the motions of the planets. *Tetrabiblos* is still the standard reference guide for the astrologer. Astronomy has been a constantly changing and advancing science, whereas astrology has essentially stood still since the days of Ptolemy, in spite of tremendous advances in our knowledge of the solar system and the universe of stars and galaxies.

A Scientist's View of Astrology

17 I continue to ask myself why people believe in astrology. I have asked this same question of many who apparently accept its predictions, including some of my young students in beginning-astronomy courses. One answer is simple and straightforward: "It would be nice to know what the future holds; so why not consult astrological predictions?" In addition, some people feel that it is useful to have available certain impersonal rules by which to make personal decisions. Astrology does provide reasonably definite answers and does yield firm guidelines for personal decisions. Many people find this very comforting indeed.

18 Believers in astrology have a remarkable faculty for remembering the times when predictions come true and ignoring the occasions when the opposite is the case. And when failure of a prediction does stare us in the face, the astrologer who made the prediction can always get out of trouble by citing the famous dictum that the stars *incline* but do not *compel.*

19 I have learned that many people who take astrology seriously were first attracted to the field by their reading of the regular columns in the newspapers. It is deplorable that so many newspapers now print this daily nonsense. At the start the regular reading is sort of a fun game, but it often ends up as a mighty serious business. The steady and ready availability of astrological predictions can over many years have insidious influences on a person's personal judgment.

20 For some people astrology has become a religion. I urge them to examine their beliefs with care. At best, astrology can be looked upon as a self-centered approach to religious beliefs, for it deals primarily with daily affairs and with what is best for a particular person. Astrology, when practiced as completely as possible, takes away from each of us our right and duty to make our own personal decisions.

21 The most complete religious approach is found in people who have "experienced" astrology, who deep inside themselves "know" astrology to be true and who believe profoundly in effects of cosmic rhythms and "vibrations." I do not know how to convince these people that they are on the wrong track, and hence they will have to go their chosen ways.

22 Many believers in astrology speak glibly of the forces exerted by the sun, moon, and planets. I should mention here that these forces—according to astrology, critically effective only at the precise moment

of birth—can hardly be gravitational or radiative in nature. The known forces that the planets exert on a child at the time of birth are unbelievably small. The gravitational forces at birth produced by the doctor and nurse and by the furniture in the delivery room far outweigh the celestial forces. And the stars are so far away from the sun and earth that their gravitational, magnetic, and other effects are negligible. Radiative effects are sometimes suggested as doing the job. First of all, the walls of the delivery room shield us effectively from many known radiations. And, second, we should bear in mind that our sun is a constantly varying source of radiation, radiating at many different wavelength variations that are by themselves far in excess of the radiation received from the moon and all the planets together.

23 Many believers in astrology have suggested that each planet issues a different variety of special, as-yet-undetected radiations or "vibrations" and that it is the interplay between these mysterious forces, or quantities, that produces strong effects of an astrological nature. If there is one thing that we have learned over the past fifty years, it is that there is apparently conclusive evidence that the sun, moon, planets, and stars are all made of the same stuff, varieties and combinations of atomic particles and molecules, all governed by uniform laws of physics. We have seen samples of the moon that are similar to rocks on earth, and as a result of our space probes we have been able to study the properties of samples from the surface of Mars. It seems inconceivable that Mars and the moon could produce mysterious waves, or vibrations, that could affect our personalities in completely different ways. It does not make sense to suppose that the various planets and the moon, all with rather similar physical properties, could manage to affect human affairs in totally dissimilar fashions.

24 There are many other questions that we can ask of the astrologers. For example, why should the precise moment of birth be *the* critical instant in a person's life? Is the instant of conception not basically a more drastic event than the precise moment when the umbilical cord is severed? Would one not expect to find in human beings the same cumulative effects that we associate with growth and environment in plants and animals? Astrology demands the existence of totally unimaginable mechanisms of force and action.

25 I shall not deal here with statistical tests of astrological predictions or with correlations. . . . At one time I thought seriously of becoming personally involved in statistical tests of astrological predic-

tions, but I abandoned this plan as a waste of time unless someone could first show me that there was some sort of physical foundation for astrology.

26 What specifically can astronomers and scientists in related fields do to make people realize that astrology is totally lacking in a proper scientific foundation? Speaking out firmly whenever the occasion demands is one way to approach the problem. This is the course that I have steadily pursued, and I hope that astronomers, young and old, will follow me on this path. I have frequently recommended that there be one or two lectures on astrology, somewhat along the lines of this essay, in each introductory astronomy course. The students should feel free to bring their questions to the instructor. An interesting experiment has been undertaken in the Natural Science course at Harvard University, in which the instructor, Michael Zeilik, is teaching a section called "Astrology—The Space Age Science?" that involves among other things a laboratory exercise in which natal horoscopes are cast.

27 The fact that some recent textbooks on astronomy contain sections (or chapters) on astrology is a most encouraging development, as is a chapter on the subject in George Abell's book *Exploring the Universe* (third edition, 1974). I have read a similar chapter in an introductory astronomy textbook now in preparation. Let us have more of this!

The Psychology of Belief in Astrology

28 Thirty-five years ago, my good friend and colleague at Harvard University, the late Gordon W. Allport, one of the finest psychologists of his day, drafted at my request a brief statement entitled "Psychologists State Their Views on Astrology." The executive council of the Society for Psychological Study of Social Issues endorsed this statement, which was publicly released, and I wish to close my present essay by reproducing it once again.

> Psychologists find no evidence that astrology is of any value whatsoever as an indicator of past, present, or future trends in one's personal life or in one's destiny. Nor is there the slightest ground for believing that social events can be foretold by divinations of the stars. The Society for the Psychological Study of Social Issues therefore deplores the faith of a considerable section of the American public in a magical practice that has no shred of justification in scientific fact.
>
> The principal reason why people turn to astrology and to kindred

superstitions is that they lack in their own lives the resources necessary to solve serious personal problems confronting them. Feeling blocked and bewildered they yield to the pleasant suggestion that a golden key is at hand—a simple solution—an ever-present help in time of trouble. This belief is more readily accepted in times of disruption and crisis when the individual's normal safeguards against gullibility are broken down. When moral habits are weakened by depression or war, bewilderment increases, self-reliance is lessened, and belief in the occult increases.

Faith in astrology or in any other occult practice is harmful insofar as it encourages an unwholesome flight from the persistent problems of real life. Although it is human enough to try to escape from the effort involved in hard thinking and to evade taking responsibility for one's own acts, it does no good to turn to magic and mystery in order to escape misery. Other solutions must be found by people who suffer from the frustrations of poverty, from grief at the death of a loved one, or from fear of economic or personal insecurity.

By offering the public the horoscope as a substitute for honest and sustained thinking, astrologers have been guilty of playing upon the human tendency to take easy rather than difficult paths. Astrologers have done this in spite of the fact that science has denied their claims and in spite of laws in some states forbidding the prophecies of astrology as fraudulent. It is against public interests for astrologers to spread their counsels of flight from reality.

It is unfortunate that in the minds of many people astrology is confused with true science. The result of this confusion is to prevent these people from developing truly scientific habits of thought that would help them understand the natural, social, and psychological factors that are actually influencing their destinies. It is, of course, true that science itself is a long way from a final solution to the social and psychological problems that perplex mankind; but its accomplishments to date clearly indicate that men's destinies are shaped by their own actions in this world. The heavenly bodies may safely be left out of account. Our fates rest not in our stars but in ourselves.

Suggestions for Discussion and Writing

1. Why does Bok, a professional astronomer, feel it is his professional obligation to debunk astrology? In his opinion, what is wrong with believing in astrology?

2. When this essay appeared in *The Humanist*, the magazine received many protesting replies. One said, in part,

"Astrology has been known for at least 4000 years. Science is about 170 years old. Which is the upstart?" If you were Bok, how would you answer this charge?

3. What facets of astrology might lead people to believe it is a science? What facets of astrology, according to Bok, make it a pseudoscience? If you were asked to advise the editor of your collegiate newspaper, would you recommend that horoscopes be published? On what grounds?

4. How would you describe the tone of Bok's essay? What words and phrases give you the best indication of his tone? What effect does the tone have on you as a reader?

5. One of the biggest scandals concerning astrology was the recent revelation that some of Ronald Reagan's actions as president were based on the advice of an astrologer. Would you vote for a candidate who followed the advice of astrologers? How is that different from following the advice of economists or pollsters?

Alfred Meyer

Do Lie Detectors Lie? All Too Often

ALFRED MEYER has published over fifty articles in the last decade, covering subjects as diverse as golden retrievers and the death of John Fitzgerald Kennedy. Some of his recent articles have dealt with science on television, brain waves, and flood control in Venice; they have appeared in *Science*, *Smithsonian*, *American Health*, and *Psychology Today*. He is also the editor of *Encountering the Environment* (1971). Meyer is a passionate environmentalist, and writes frequently about the subject in his role as Editor-at-Large for *Mother Earth News*. This essay first appeared in *Science* in 1982.

About the size of an attaché case and equally portable, the instrument looks fairly innocuous. Dials and knobs cluster neatly at one end, while four inked styluses rest on a roll of paper on the other. Once the tubes, cuffs, and electrodes that are its sensors emerge, however, the instrument appears more sinister, doubly so when a human subject is connected to its serpentine appurtenances. But the hardware inflicts no injury, merely records changes in some rather simple physiological responses: blood pressure, breathing, sweating. On the contrary, it is the software programmed into the instrument's keeper, a sort of wizard of truth, that is menacing. He will be well-dressed, efficient, and maddeningly neutral. In calm tones he will tell you how the instrument works and assure you that it is highly accurate. He will discuss the questions he intends to ask, switch on the instrument, and begin the questioning. Within an hour or two, he will leave, in most cases convinced that, by combining his own observations of your behavior with the readings of his instrument, he knows if you have told the truth.

2 Limited almost exclusively to law enforcement work and to matters of national security during its formative years in the 1940s and 1950s, the lie detector, or polygraph, today has become a common fixture in American society, almost to an Orwellian degree. In addition to its uses in prisons, the military, police work, FBI and CIA investigations, and pretrial examinations both for the prosecution and the defense, the polygraph has also found its way into corporate America, where it is widely used for detecting white collar crime and for screening potential employees. This year, it is estimated, half a million to a million Americans, for one reason or another, will take a lie detector test. Such extravagant use takes place despite the facts that no state freely admits polygraph evidence in criminal cases (though about 25 do when both prosecution and defense agree in advance of the testing), that innocent persons often fail the polygraph test while guilty ones pass it, and that the scientific foundation of polygraph technology is, if not altogether questionable, as some critics claim, at least open to serious challenge.

3 Ironically, considering the procedure's inherent piety, the administration of the lie detector test involves a necessary touch of deception. Polygraph theory holds that physiological reactions—changes in blood pressure, rate of breathing, sweating of the palms—elicited by a set of questions will reliably betray falsehood. But it is

the form and mix of questions that polygraphers claim are the key to their technique. The standard format, known as the Control Question Test, involves interspersing "relevant" questions with "control" questions. Relevant questions relate directly to the critical matter, such as, "Did you participate in the robbery of the First National Bank on September 11, 1981?" Control questions, on the other hand, are less precise, such as, "In the last 20 years, have you ever taken something that didn't belong to you?"

4 In the pretest interview, the polygrapher reviews all the questions and frames the control questions to produce "no" answers. It is in this crucial pretest phase that the polygrapher's deception comes into play, for he wants the innocent subject to dissemble while answering the control questions during the actual test. For example, pursuing the hypothetical control question above, most people are likely to have taken something that didn't belong to them at some stage in their lives. Yet fear of embarrassment may lead the subjects to deny misdeeds and therefore to answer the control question dishonestly.

5 The assumption underlying the Control Question Test is that the truthful subject will display a stronger physiological reaction to the control questions, whereas a deceptive subject will react more strongly to the relevant questions.

6 That is the heart of it. Modern lie detection relies on nothing more than subtle psychological techniques, crude physiological indicators, and skilled questioning and interpretation of the results. It resembles nothing so much as a game of cat and mouse—often played for very high stakes.

7 But does it work? The answer is, yes, sometimes. As for how well—that depends on who you ask.

8 In a recent publication, for example, the American Polygraph Association cites studies that yield accuracy rates ranging from 87.2 to 96.2 percent. But such studies, conducted in the lab by criminologists and a handful of psychologists, most of whom make their living as polygraphers, have come under heavy attack. The chief adversary is psychologist David Lykken of the University of Minnesota, who cites other studies, conducted in the field, which yield far lower rates of 64 to 71 percent.

9 Lykken bases his criticism of lab data on several points. First, he maintains that there is no "specific lie response." The polygraph merely records general emotional arousal. It cannot distinguish anxi-

ety or indignation from guilt. Second—and this is his most telling point—accuracy rates based on lab studies are flawed because they depend on mock crimes using subjects who do not face the real life consequences of being found truthful or deceptive.

10 Even a stalwart of polygraphy like psychophysiologist David Raskin of the University of Utah acknowledges the difficulty of assessing accuracy rates in real life situations, those where people face criminal charges. Moreover, he writes, "In a large percentage of criminal investigations, guilt or innocence is never determined conclusively."

11 Even in field trials, for example, where a panel of polygraph experts independently score the tests of people who have previously been tried and found guilty or not guilty, one can't be sure the outcome is accurate. Such ambiguity makes it virtually impossible to correlate polygraph results with the truth in samples large enough to achieve genuine statistical validity.

12 When it comes to verifying the truthfulness of innocent subjects, both Raskin and Lykken agree that the lie detector turns up more innocent people found guilty—false positives—than guilty people found innocent. While Raskin believes the chances of a polygraph identifying an innocent person as lying are less than 10 percent, Lykken believes that the chance is closer to 50 percent. He contends that it is impossible to design control questions that will produce the same level of responsiveness in innocent subjects as is produced by the relevant questions in guilty subjects. The discrepancy results from lack of agreement on whether to use lab or field studies as the standard.

13 Lykken has still other criticisms of the polygraph. He contends, for example, that with some preparation it is possible to defeat the test. To illustrate, he frequently relates the experience of Floyd Fay, convicted of a murder charge—wrongfully, it was later proved—partly on the basis of a failed lie detector test. While imprisoned, Fay took an interest in polygraph technology and, after studying it, claimed to have trained fellow inmates to foil tests administered by prison officials. The inmates reportedly used such countermeasures as placing a tack in a shoe and stamping on it during the control questioning.

14 Raskin dismisses Fay's training efforts since they were hardly run under controlled conditions. He cites the recent work of psychologist Michael Dawson in California. With the help of the late

method actor Lee Strasberg, Dawson invited a large group of students from Strasberg's Actors' Studio in Hollywood to try, with all an actor's cunning control of emotions, à la Stanislavsky, to "beat the test." None did. Yet again, it was an experiment involving a mock rather than a real crime.

15 Clearly no one has established beyond a doubt the validity and reliability of the lie detector, though many argue that it is better than any of the alternatives, such as expert testimony from psychiatrists or psychologists. Yet polygraphy is fast becoming an American obsession, one, incidentally, not shared by the British or the Europeans or, as far as we know, the Russians. America's increasing dependence on the polygraph reflects its enormous faith in the rational processes of science: Each of us can probably recall a time when our voices sounded funny as we told a fib. Surely, if we can "hear" a lie, science can detect one.

16 It comes as a surprise, therefore, to learn how fragile polygraphy's scientific foundations are. When Raskin embarked on polygraphy research in Utah in 1970, not a single scientific laboratory had assessed the validity or reliability of the Control Question Test, by then already in wide use. Nor had a single report by a trained scientist evaluating that test appeared in the scientific literature.

17 Meanwhile, armed rather more with art than with science—and not a little merchandising magic thrown in—the technology is spreading like wildfire, alarming even such advocates as Raskin. Though he and Lykken clash often enough in print on the scientific merits of polygraphy, there is one development that appalls both men equally: use of polygraph tests in pre-employment screening by corporations, banks, fast-food chains, and a number of other commercial enterprises.

18 Such tests, according to Raskin, often are conducted hastily and haphazardly, resulting in highly questionable accuracy. They seek by vague and general questions to elicit admissions of a personal nature, thereby constituting invasion of privacy, and violate personal freedom by requiring tests that may put a job at stake. Particularly distasteful are tests given to present employees—who can hardly refuse—such as those recently used in the upper echelons of government to try to curb leaks to the media.

19 For Lykken, such uses merely exacerbate an already intolerable situation. He points to the scant amount of training polygraphers re-

ceive before they qualify as technicians of truth. Perhaps as many as a dozen contemporary polygraphers do hold Ph.D. degrees, but the vast majority of the 4,000 to 8,000 practicing examiners have had no significant training in physiology or in psychology, even though lie detection demands extremely subtle—and difficult—psychophysiological interpretations. Most of the 25 or more schools that train examiners provide only an eight-week course of instruction and require two years of college for admission.

20 While Raskin in his laboratory continues to attempt to make the polygraph more scientifically valid and reliable, Lykken continues to question the ground rules. Where Raskin sees the polygraph as a legitimate tool for protecting society, Lykken fears that the technology poses a greater threat to the innocent than to the guilty and, ultimately, is based on little more than an intimidating mystique that uses the language and trappings of science to sell itself to a trusting public.

21 Most law enforcement officials, however, look on the polygraph not as a technique for intimidating and imprisoning people but for screening out innocent suspects early, thus expediting the legal process. They prefer to talk about truth verification rather than lie detection. But for the present, verifying the truth appears at least as difficult as detecting a lie.

Suggestions for Discussion and Writing

1. Is it appropriate to call the polygraph machine a "lie detector"? What, according to Meyer, can the polygraph detect?

2. What are the major objections to the use of polygraph results? Can people "beat" the polygraph? Under what conditions?

3. Meyer lines up a number of experts to "give testimony" as he reviews the use of polygraphs. How does he integrate this testimony into his argument? What effect does this "expert testimony" have on you as readers?

4. Suppose you were on a jury and your decision to convict the defendant in a major case came down to whether you believed evidence obtained from a polygraph test. Could you vote to convict that defendant? Would Meyer advise you to do so?

5. How widespread is the use of polygraphs in government and business today? Do you find any evidence that "polygraphy is fast becoming an obsession," as Meyer claims?

3 Scientists "Read" Nature

"Nature is a textbook for those who can read it," the great Harvard biologist Louis Agassiz once wrote. For many scientists, nature is the textbook they study all their lives; it never goes out of date, nor requires a new edition, though the "answers" to the questions nature poses change constantly. But nature is also a textbook written by humanity, as scientists use their perceptions and observations to "read" nature, and to discover things about humanity through that reading. They don't create nature; rather, they create new interpretations of what they have learned.

In this section we present five examples of "reading" nature, all leading to different conclusions. Stephen Jay Gould considers the concept of "natural time," and explores the misunderstandings people have about it. Alexander Petrunkevitch finds in insect behaviors a lesson for people. Pamela Weintraub summarizes attempts to understand the nature of human brains, and Barry Lopez describes the interdependence of humans and other species. Finally, Loren Eiseley explores the place of awe and respect in scientific study. As you read these essays, ask yourself, "What is nature? Is it the opposite of science? the partner of it? How do I regard nature? Why do I regard it that way?"

Stephen Jay Gould

Our Allotted Lifetimes

Harvard University professor STEPHEN JAY GOULD (born 1941) has won praise for his ability to make difficult scientific theories understandable by linking science with ideas from literature, art, music, and many other of the humanities. "Science is not a heartless pursuit of objective information," he has said, "it is a creative human activity." With Niles Eldredge, Gould is a leading proponent of "punctuated equilibrium," the idea that evolution occurs not gradually and steadily, but suddenly in small splinter groups of a species.

Although Gould says he doesn't describe himself as a writer, he has published steadily, including his monthly column in *Natural History* magazine. He won the American Book Award for *The Panda's Thumb: More Reflections in Natural History* (1981), and the National Book Critics' Award for *The Mismeasure of Man* (1981). Gould says, "I've always written very intuitively. I was never trained in writing. . . . What I've developed is a style that more or less grew with my practice of it." Gould's most recent book is *Bully for Brontosaurus* (1991). This essay first appeared in *Natural History* in 1977.

Meeting with Henry Ford in E. L. Doctorow's *Ragtime*, J. P. Morgan praises the assembly line as a faithful translation of nature's wisdom:

> Has it occurred to you that your assembly line is not merely a stroke of industrial genius but a projection of organic truth? After all, the interchangeability of parts is a rule of nature. . . . All mammals reproduce in the same way and share the same designs of self-nourishment, with digestive and circulatory systems that are recognizably

> the same, and they enjoy the same senses. . . . Shared design is what allows taxonomists to classify mammals as mammals.

2 An imperious tycoon should not be met with equivocation; nonetheless, I can only reply "yes, and no" to Morgan's pronouncement. Morgan was wrong if he thought that large mammals are geometric replicas of small ones. Elephants have relatively smaller brains and thicker legs than mice, and these differences record a general rule of mammalian design, not the idiosyncracies of particular animals.

3 Morgan was right in arguing that large animals are essentially similar to small members of their group. The similarity, however, does not lie in a constant shape. The basic laws of geometry dictate that animals must change their shape in order to perform the same function at different sizes. I remind readers of the classical example, first discussed by Galileo in 1638: the strength of an animal's leg is a function of its cross-sectional area (length × length); the weight that the leg must support varies as the animal's volume (length × length × length). If a mammal did not alter the relative thickness of its legs as it got larger, it would soon collapse since body weight would increase much faster than the supporting strength of limbs. Instead, large mammals have relatively thicker leg bones than small mammals. To remain the same in function, animals must change their form.

4 The study of these changes in form is called "scaling theory." Scaling theory has uncovered a remarkable regularity of changing shape over the 25-millionfold range of mammalian weight from shrew to blue whale. If we plot brain weight versus body weight for all mammals on the so-called mouse-to-elephant (or shrew-to-whale) curve, very few species deviate far from a single line expressing the general rule: brain weight increases only two-thirds as fast as body weight as we move from small to large mammals. (We share with bottle-nosed dolphins the honor of greatest deviance from the curve.)

5 We can often predict these regularities from the physical behavior of objects. The heart, for example, is a pump. Since all mammalian hearts are similar in function, small hearts will pump considerably faster than large ones (imagine how much faster you could work a finger-sized toy bellows than the giant model that fuels a blacksmith's large forge). On the mouse-to-elephant curve for mammals, the length of a heartbeat increases between one-fourth and one-third as fast as body weight as we move from small to large mammals.

The generality of this conclusion has just been affirmed in an interesting study by J. E. Carrel and R. D. Heathcote on the scaling of heart rate in spiders. They used a cool laser beam to illuminate the hearts of resting spiders and drew a crab spider-to-tarantula curve for eighteen species spanning nearly a thousandfold range of body weight. Again, scaling is very regular with heart rate increasing four-tenths as fast as body weight (or .409 times as fast, to be exact).

6 We may extend this conclusion for hearts to a very general statement about the pace of life in small versus large animals. Small animals tick through life far more rapidly than large ones—their hearts work more quickly, they breathe more frequently, their pulse beats much faster. Most importantly, metabolic rate, the so-called fire of life, scales only three-fourths as fast as body weight in mammals. Large mammals generate much less heat per unit of body weight to keep themselves going. Tiny shrews move frenetically, eating nearly all their waking lives to keep their metabolic fire burning at its maximal rate among mammals; blue whales glide majestically, their hearts beating the slowest rhythm among active, warmblooded creatures.

7 If we consider the scaling of lifetime among mammals, an intriguing synthesis of these disparate data seems to suggest itself. We have all had enough experience with mammalian pets of various sizes to understand that small mammals tend to live for a shorter time than large ones. In fact, the scaling of mammalian lifetime follows a regular curve at about the same rate as heartbeat and breath time—between one-fourth and one-third as fast as body weight as we move from small to large animals. (Again, *Homo sapiens* emerges as a very peculiar animal. We live far longer than a mammal of our body size should. I have argued elsewhere that humans evolved by a process called "neoteny"—the retention of shapes and growth rates that characterize juvenile stages of our primate ancestors. I also believe that neoteny is responsible for our elevated longevity. Compared with other mammals, all stages of human life—from juvenile features to adulthood—arise "too late." We are born as helpless embryos after a long gestation; we mature late after an extended childhood; we die, if fortune be kind, at ages otherwise reached only by the very largest warmblooded creatures.)

8 Usually, we pity the pet mouse or gerbil that lived its full span of a year or two at most. How brief its life, while we endure for the

better part of a century. As the main theme of this column, I want to argue that such pity is misplaced (our personal grief, of course, is quite another matter; with this, science does not deal). J. P. Morgan of *Ragtime* was right—small and large mammals are essentially similar. Their lifetimes are scaled to their life's pace, and all endure for approximately the same amount of biological time. Small mammals tick fast, burn rapidly, and live for a short time; large ones live long at a stately pace. Measured by their own internal clocks, mammals of different sizes tend to live for the same amount of time.

9 Yet we are prevented from grasping this important and comforting concept by a deeply ingrained habit of Western thought. We are trained from earliest memory to regard absolute Newtonian time as the single valid measuring stick in a rational and objective world. We impose our kitchen clock, ticking equably, upon all things. We marvel at the quickness of a mouse, express boredom at the torpor of a hippopotamus. Yet each is living at the appropriate pace of its own biological clock.

10 I do not wish to deny the importance of absolute, astronomical time to organisms. Animals must measure it to lead successful lives. Deer must know when to regrow their antlers, birds when to migrate. Animals track the day-night cycle with their circadian rhythms; jet lag is the price we pay for moving much faster than nature intended. Bamboos can somehow count 120 years before flowering again.

11 But absolute time is not the appropriate measuring stick for all biological phenomena. Consider the song of the humpback whale. These magnificent animals sing with such volume that their sounds travel through water for thousands of miles, perhaps even around the world, as their leading student Roger S. Payne has suggested. E. O. Wilson has described the awesome effect of these vocalizations: "The notes are eerie yet beautiful to the human ear. Deep basso groans and almost inaudibly high soprano squeaks alternate with repetitive squeals that suddenly rise or fall in pitch." We do not know the function of these songs. Perhaps they enable whales to find each other and to stay together during their annual transoceanic migrations.

12 Each whale has its own characteristic song; the highly complex patterns are repeated over and over again with great faithfulness. No scientific fact that I have learned in the last decade struck me with more force than Payne's report that the length of some songs may extend for more than half an hour. I have never been able to memorize

the five-minute first Kyrie of the B minor Mass (and not for want of trying); how could a whale sing for thirty minutes and then repeat itself accurately? Of what possible use is a thirty-minute repeat cycle—far too long for a human to recognize: we would never grasp it as a single song (without Payne's recording machinery and much study after the fact). But then I remembered the whale's metabolic rate, the enormously slow pace of its life compared with ours. What do we know about a whale's perception of thirty minutes? A humpback may scale the world to its own metabolic rate: its half-hour song may be our minute waltz.2 From any point of view, the song is spectacular; it is the most elaborate single display so far discovered in any animal. I merely urge the whale's point of view as an appropriate perspective.

13 We can provide some numerical precision to support the claim that all mammals, on average, live for the same amount of biological time. In a method developed by W. R. Stahl, B. Gunther, and E. Guerra in the late 1950s and early 1960s, we search the mouse-to-elephant equations for biological properties that scale at the same rate against body weight. For example, Gunther and Guerra give the following equations for mammalian breath time and heartbeat time versus body weight.

$$\text{breath time} = .0000470\ \text{body}^{0.28}$$
$$\text{heartbeat time} = .0000119\ \text{body}^{0.28}$$

(Nonmathematical readers need not be overwhelmed by the formalism. The equations simply mean that both breath time and heartbeat time increase about .28 times as fast as body weight as we move from small to large mammals.) If we divide the two equations, body weight cancels out because it is raised to the same power.

$$\frac{\text{breath time}}{\text{heartbeat time}} = \frac{.0000470\ \cancel{\text{body}^{0.28}}}{.0000119\ \cancel{\text{body}^{0.28}}} = 4.0$$

14 This says that the ratio of breath time to heartbeat time is 4.0 in mammals of any body size. In other words, all mammals, whatever their size, breathe once for each four heartbeats. Small animals breathe and beat their hearts faster than large animals, but both breath and heart slow up at the same relative rate as mammals get larger.

15 Lifetime also scales at the same rate to body weight (.28 times as fast as we move from small to large mammals). This means that the ratio of both breath time and heartbeat time to lifetime is also constant over the whole range of mammalian size. When we perform an exercise similar to that above, we find that all mammals, regardless of their size, tend to breathe about 200 million times during their lives (their hearts, therefore, beat about 800 million times). Small mammals breathe fast, but live for a short time. Measured by the sensible internal clocks of their own hearts or the rhythm of their own breathing, all mammals live about the same time. (Astute readers, having counted their breaths, may have calculated that they should have died long ago. But *Homo sapiens* is a markedly deviant mammal in more ways than braininess alone. We live about three times as long as mammals of our body size "should," but we breathe at the "right" rate and thus live to breathe about three times as much as an average mammal of our body size.)

16 The mayfly lives but a day as an adult. It may, for all I know, experience that day as we live a lifetime. Yet all is not relative in our world, and such a short glimpse of it must invite distortion in interpreting events ticking on longer scales. In a brilliant metaphor, the pre-Darwinian evolutionist Robert Chambers spoke of a mayfly watching the metamorphosis of a tadpole into a frog (from *Vestiges of the Natural History of Creation*, 1844):

> Suppose that an ephemeron [a mayfly], hovering over a pool for its one April day of life, were capable of observing the fry of the frog in the waters below. In its aged afternoon, having seen no change upon them for such a long time, it would be little qualified to conceive that the external branchiae [gills] of these creatures were to decay, and be replaced by internal lungs, that feet were to be developed, the tail erased, and the animal then to become a denizen of the land.

17 Human consciousness arose but a minute before midnight on the geologic clock. Yet we mayflies, ignorant perhaps of the messages buried in earth's long history, try to bend an ancient world to our purposes. Let us hope that we are still in the morning of our April day.

Suggestions for Discussion and Writing

1. Why does Gould think it's important for people to understand scaling theory? What sorts of changes in our thinking

should an understanding of scaling theory lead to?

2. We're often told that one human year of a dog's life is equal to seven dog years. How would Gould's explanation of scaling theory explain the comparative ages of dogs and people?

3. What is Gould's thesis in this essay? Where does he state it? What clues does he provide to lead readers to this thesis?

4. What is a metaphor? What is the purpose of the mayfly metaphor that Gould uses in his conclusion?

5. What is your perception of the role time plays in your life? Does time really fly? What does it mean not "to waste time"? How would Gould suggest you revise your opinion of time?

Alexander Petrunkevitch

The Spider and the Wasp

ALEXANDER IVANOVICH PETRUNKEVITCH was born in the Ukraine in 1875, but left there after his youthful political activities made him an "undesirable" in Tsarist Russia. He received a Ph.D. at Freiburg in 1900, emigrated to America, and taught at Harvard before becoming a professor of zoology at Yale from 1917-1944. There, although his research also included studies in heredity and psychology, he became known as "the man who knows more than anyone else in the world about spiders." He wrote what is still one of the world's most comprehensive catalogs of spiders.

"Pete", as his friends called him, believed in intense research. In a 1951 interview, he estimated that he had spent over forty thousand hours in collecting spiders for study, and another two hundred thousand on research, teaching, writing, and other academic activities. "You cannot understand any machine unless you know how it's built," he would tell his students. "Likewise, you must know an animal's morphology to know its true nature." He remained active in the scientific community until his death in 1964. This essay was first published in 1952.

In the feeding and safeguarding of their progeny insects and spiders exhibit some interesting analogies to reasoning and some crass examples of blind instinct. The case I propose to describe here is that of the tarantula spiders and their archenemy, the digger wasps of the genus *Pepsis.* It is a classic example of what looks like intelligence pitted against instinct—a strange situation in which the victim, though fully able to defend itself, submits unwittingly to its destruction.

2 Most tarantulas live in the tropics, but several species occur in the temperate zone and a few are common in the southern U.S. Some varieties are large and have powerful fangs with which they can inflict a deep wound. These formidable-looking spiders do not, however, attack man; you can hold one in your hand, if you are gentle, without being bitten. Their bite is dangerous only to insects and small mammals such as mice; for man it is no worse than a hornet's sting.

3 Tarantulas customarily live in deep cylindrical burrows, from which they emerge at dusk and into which they retire at dawn. Mature males wander about after dark in search of females and occasionally stray into houses. After mating, the male dies in a few weeks, but a female lives much longer and can mate several years in succession. In a Paris museum is a tropical specimen which is said to have been living in captivity for 25 years.

4 A fertilized female tarantula lays from 200 to 400 eggs at a time; thus it is possible for a single tarantula to produce several thousand young. She takes no care of them beyond weaving a cocoon of silk to enclose the eggs. After they hatch, the young walk away, find convenient places in which to dig their burrows and spend the rest of their lives in solitude. The eyesight of tarantulas is poor, being limited to a sensing of change in the intensity of light and to the perception of moving objects. They apparently have little or no sense of hearing, for a hungry tarantula will pay no attention to a loudly chirping cricket

placed in its cage unless the insect happens to touch one of its legs.

5 But all spiders, and especially hairy ones, have an extremely delicate sense of touch. Laboratory experiments prove that tarantulas can distinguish three types of touch: pressure against the body wall, stroking of the body hair, and riffling of certain very fine hairs on the legs called trichobothria. Pressure against the body, by the finger or the end of a pencil, causes the tarantula to move off slowly for a short distance. The touch excites no defensive response unless the approach is from above where the spider can see the motion, in which case it rises on its hind legs, lifts its front legs, opens its fangs and holds this threatening posture as long as the object continues to move.

6 The entire body of a tarantula, especially its legs, is thickly clothed with hair. Some of it is short and wooly, some long and stiff. Touching this body hair produces one of two distinct reactions. When the spider is hungry, it responds with an immediate and swift attack. At the touch of a cricket's antennae the tarantula seizes the insect so swiftly that a motion picture taken at the rate of 64 frames per second shows only the result and not the process of capture. But when the spider is not hungry, the stimulation of its hairs merely causes it to shake the touched limb. An insect can walk under its hairy belly unharmed.

7 The trichobothria, very fine hairs growing from dislike membranes on the legs, are sensitive only to air movement. A light breeze makes them vibrate slowly, without disturbing the common hair. When one blows gently on the trichobothria, the tarantula reacts with a quick jerk of its four front legs. If the front and hind legs are stimulated at the same time, the spider makes a sudden jump. This reaction is quite independent of the state of its appetite.

8 These three tactile responses—to pressure on the body wall, to moving of the common hair, and to flexing of the trichobothria—are so different from one another that there is no possibility of confusing them. They serve the tarantula adequately for most of its needs and enable it to avoid most annoyances and dangers. But they fail the spider completely when it meets its deadly enemy, the digger wasp *Pepsis.*

9 These solitary wasps are beautiful and formidable creatures. Most species are either a deep shiny blue all over, or deep blue with rusty wings. The largest have a wing span of about four inches. They live on nectar. When excited, they give off a pungent odor—a warning that they are ready to attack. The sting is much worse than that of a bee or common wasp, and the pain and swelling last longer. In the adult

stage the wasp lives only a few months. The female produces but a few eggs, one at a time at intervals of two or three days. For each egg the mother must provide one adult tarantula, alive but paralyzed. The mother wasp attaches the egg to the paralyzed spider's abdomen. Upon hatching from the egg, the larva is many hundreds of times smaller than its living but helpless victim. It eats no other food and drinks no water. By the time it has finished its single Gargantuan meal and become ready for wasphood, nothing remains of the tarantula but its indigestible chitinous skeleton.

10 The mother wasp goes tarantula-hunting when the egg in her ovary is almost ready to be laid. Flying low over the ground late on a sunny afternoon, the wasp looks for its victim or for the mouth of a tarantula burrow, a round hole edged by a bit of silk. The sex of the spider makes no difference, but the mother is highly discriminating as to species. Each species of *Pepsis* requires a certain species of tarantula, and the wasp will not attack the wrong species. In a cage with a tarantula which is not its normal prey, the wasp avoids the spider and is usually killed by it in the night.

11 Yet when a wasp finds the correct species, it is the other way about. To identify the species the wasp apparently must explore the spider with her antennae. The tarantula shows an amazing tolerance to this exploration. The wasp crawls under it and walks over it without evoking any hostile response. The molestation is so great and so persistent that the tarantula often rises on all eight legs, as if it were on stilts. It may stand this way for several minutes. Meanwhile the wasp, having satisfied itself that the victim is of the right species, moves off a few inches to dig the spider's grave. Working vigorously with legs and jaws, it excavates a hole 8 to 10 inches deep with a diameter slightly larger than the spider's girth. Now and again the wasp pops out of the hole to make sure that the spider is still there.

12 When the grave is finished, the wasp returns to the tarantula to complete her ghastly enterprise. First she feels it all over once more with her antennae. Then her behavior becomes more aggressive. She bends her abdomen, protruding her sting, and searches for the soft membrane at the point where the spider's legs join its body—the only spot where she can penetrate the horny skeleton. From time to time, as the exasperated spider slowly shifts ground, the wasp turns on her back and slides along with the aid of her wings, trying to get under the tarantula for a shot at the vital spot. During all this maneuvering,

which can last for several minutes, the tarantula makes no move to save itself. Finally the wasp corners it against some obstruction and grasps one of its legs in her powerful jaws. Now at last the harassed spider tries a desperate but vain defense. The two contestants roll over and over on the ground. It is a terrifying sight and the outcome is always the same. The wasp finally manages to thrust her sting into the soft spot and holds it there for a few seconds while she pumps in the poison. Almost immediately the tarantula falls paralyzed on its back. Its legs stop twitching; its heart stops beating. Yet it is not dead, as is shown by the fact that if taken from the wasp it can be restored to some sensitivity by being kept in a moist chamber for several months.

13 After paralyzing the tarantula, the wasp cleans herself by dragging her body along the ground and rubbing her feet, sucks a drop of blood oozing from the wound in the spider's abdomen, then grabs a leg of the flabby, helpless animal in her jaws and drags it down to the bottom of the grave. She stays there for many minutes, sometimes for several hours, and what she does all that time in the dark we do not know. Eventually she lays her egg and attaches it to the side of the spider's abdomen with a sticky secretion. Then she emerges, fills the grave with soil carried bit by bit in her jaws, and finally tramples the ground all around to hide any trace of the grave from prowlers. Then she flies away, leaving her descendant safely started in life.

14 In all this the behavior of the wasp evidently is qualitatively different from that of the spider. The wasp acts like an intelligent animal. This is not to say that instinct plays no part or that she reasons as man does. But her actions are to the point; they are not automatic and can be modified to fit the situation. We do not know for certain how she identifies the tarantula—probably it is by some olfactory or chemo-tactile sense—but she does it purposefully and does not blindly tackle a wrong species.

15 On the other hand, the tarantula's behavior shows only confusion. Evidently the wasp's pawing gives it no pleasure, for it tries to move away. That the wasp is not simulating sexual stimulation is certain because male and female tarantulas react in the same way to its advances. That the spider is not anesthetized by some odorless secretion is easily shown by blowing lightly at the tarantula and making it jump suddenly. What, then, makes the tarantula behave as stupidly as it does?

16 No clear, simple answer is available. Possibly the stimulation by the wasp's antennae is masked by a heavier pressure on the spider's

body, so that it reacts as when prodded by a pencil. But the explanation may be much more complex. Initiative in attack is not in the nature of tarantulas; most species fight only when cornered so that escape is impossible. Their inherited patterns of behavior apparently prompt them to avoid problems rather than attack them. For example, spiders always weave their webs in three dimensions, and when a spider finds that there is insufficient space to attach certain threads in the third dimension, it leaves the place and seeks another, instead of finishing the web in a single plane. This urge to escape seems to arise under all circumstances, in all phases of life, and to take the place of reasoning. For a spider to change the pattern of its web is as impossible as for an inexperienced man to build a bridge across a chasm obstructing his way.

17 In a way the instinctive urge to escape is not only easier but often more efficient than reasoning. The tarantula does exactly what is most efficient in all cases except in an encounter with a ruthless and determined attacker dependent for the existence of her own species on killing as many tarantulas as she can lay eggs. Perhaps in this case the spider follows its usual pattern of trying to escape, instead of seizing and killing the wasp, because it is not aware of its danger. In any case, the survival of the tarantula species as a whole is protected by the fact that the spider is much more fertile than the wasp.

Suggestions for Discussion and Writing

1. On the surface, this essay seems to be about bugs. Can you determine another purpose for Petrunkevitch's writing it? Does the year in which it was written give you any clue to what that other purpose is? Would you call this essay a parable? Why or why not?

2. Why does Petrunkevitch call the battle between wasps and spiders a battle between intelligence and instinct? Why does he say that an instinctive response is "not only easier but often more efficient than reasoning"?

3. The organizational method Petrunkevitch most uses here is comparison and contrast. Why do you think he chose this method?

4. The initial description of the tarantula takes up more space than that of the wasp. Why do you think Petrunkevitch chose to provide more detail about the one than the other?

5. Pick some current issue (local or national) that interests you, and write a parable about it. Try to find a set of familiar metaphoric characters (like the tortoise and the hare or the three pigs) that will help you make your point more emphatically.

Pamela Weintraub

The Brain: His and Hers

The journalist PAMELA WEINTRAUB, born in 1954, is a native of Brooklyn who graduated from SUNY-Albany with a double major in Biology and English, then went on to receive a master's degree in science journalism from Boston University. She has published articles on subjects covering a wide range: health, UFOs, sports training, in-utero teaching of babies, memory, nutrition, and genetics. Her work appears frequently in *Omni* and *Ms.* With her brother Alan Weintraub, she is also the author of *Twenty-Five Things You Can Do to Beat the Recession of the Nineties* (1991). She and Michael Fox also co-authored *Save the Animals* (1991).

Weintraub says, "One of the things I've learned about writing is that it's truly a *process* in which you continually hone things; what's important is not the first draft, but the second, and the third." This essay first appeared in *Discover* in 1981.

Are the brains of men and women different? If so, do men and women differ in abilities, talents, and deficiencies? A scientific answer to these questions could affect society and culture, and variously shock, intrigue, delight, depress, and reassure people of both sexes. Now an answer is coming into sight: Yes, male and female brains do differ.

2 That men and women think and behave differently is a widely held assumption. Generations of writers have lavished their attention on these differences, proclaiming, for example, that aggressiveness and promiscuity are natural to the male, that domesticity is the legacy of the female. Today's feminists acknowledge some differences, hut hotly dispute the notion that they are innate. They stress that it is society, not nature, that gives men the drive to dominate and keeps women from achieving careers and power. But proof that behavioral and intellectual differences between the sexes are partly rooted in the structure of the brain, that women are inherently superior in some areas of endeavor and men in others would in no way undermine legitimate demands for social equality. Instead the result could be a better, more realistic relationship between the sexes.

3 The evidence suggesting differences between male and female brains comes from research in behavior, biochemistry, anatomy, and neuropsychology. The most recent study deals with the long-established fact that skill in mathematics is far more common among men than women. Feminists—and many scientists—blame sexual stereotyping. But psychologists Camilla Benbow and Julian Stanley, at Johns Hopkins University, challenged that interpretation after testing 9,927 seventh and eighth graders with high IQs. As Benbow told *Discover* reporter John Bucher, of the students who scored 500 or better on the math part of the Scholastic Aptitude Test, boys outnumbered girls by more than two to one. In other words, the psychologists argue, male superiority in math is so pronounced that, to some extent, it must be inborn.

4 This finding follows several recent studies proving that male and female brains, at least in animals, are physically different. From the hypothalamus, the center for sexual drive, to the cerebral cortex, the seat of thought, scientists have found consistent variations between the sexes. The causes of these differences, they say, are the sex hormones—the male androgens and female estrogens and progesterones that are secreted by the sex glands and carried through the blood stream to distant parts of the body, where they control everything from menstruation to the growth of facial hair.

5 Basic to all the studies of gender and the brain are the facts of sex determination. When a child is conceived, each parent contributes a sex chromosome, either an X or a Y (so-called for their shapes). When two X's combine, the fetus develops ovaries and becomes a girl. An X

and a Y produce a boy; the Y chromosome makes a protein that coats the cells programmed to become ovaries, directing them to become testicles instead. The testicles then pump out two androgens, one that absorbs what would have become a uterus, and another, testosterone, that causes a penis to develop.

6 Though scientists have not yet been able to pinpoint any physiological differences between the brains of men and women, they think that the development of the brain parallels that of the genitals. If the fetus is a boy, they say, the testosterone that produces the penis also masculinizes tissue in the hypothalamus and other nearby structures deep within the brain. New data suggest that if the fetus is a girl, estrogen secreted by the ovaries feminizes brain tissue in the surrounding cerebral cortex. Scientists cannot dissect living human brains, but they have found ingenious ways to test their theories. The major approaches:

Human Behavior

7 To shed light on the sexuality of the brain, endocrinologist Julianne Imperato-McGinley of Cornell Medical College in New York City studied 38 men in an isolated part of the Dominican Republic who, because of a genetic disorder, started life as girls. They stayed indoors playing with dolls and learning to cook while boys fought and shouted outside. At the age of eleven, when the breasts of normal girls began to enlarge, the children studied by Imperato-McGinley showed no change. But at twelve, most of them began to feel stirrings of sexual desire for girls. At puberty, their voices deepened, their testicles descended, and their clitorises enlarged to become penises.

8 These children came from a group of families carrying a rare mutant gene that deprived them of an enzyme needed to make testosterone work in the skin of their genitals. For this reason their external genitals looked female at birth. But at puberty their bodies were able to use testosterone without the enzyme, and it became obvious that they were males—as chromosome tests confirmed. All but two are now living with women. They have male musculature and, although they cannot sire children, they can have sexual intercourse. They have assumed masculine roles in their society. "To the world," says Imperato-McGinley, "they looked like girls when they were younger. But their bodies were actually flooded with testosterone." She concludes that

they were able to adjust easily because hidden in the girl's body was a male brain, virilized by testosterone before birth and activated by another rush of testosterone during adolescence.

9 Although Imperato-McGinley suggests that brain structure determines behavior, another scientist thinks that the reverse may also be true: Anne Petersen, director of the Adolescent Laboratory at the Michael Reese Hospital and Medical Center in Chicago, says that cultural experiences can masculinize or feminize the brain. In a recent study, Petersen found that boys who excel in athletics also excel in spatial reasoning—a skill controlled by the right hemisphere of the cerebral cortex, and defined as the ability to understand maps and mazes or objects rotating in space. Says Petersen, "An athlete must be constantly aware of his own body and a whole constellation of other bodies in space." A daily game of basketball might, through some still mysterious mechanism, stimulate the secretion of hormones that prime a player's brain for success in basketball. The same brain structures would be used to deal with spatial problems. "Women are far less athletic than men," says Petersen, "and also less adept at spatial reasoning. Part of the problem may be their lack of involvement in sports. Perhaps some women just never develop the area of the brain specialized for spatial control."

10 Like Petersen, endocrinologist Anke Ehrhardt thinks that society plays an important part in shaping gender behavior. Nevertheless, she says, "certain types of sexual behavior are influenced by the sex hormones." Leafing through the clutter of papers and books that cover her desk at New York City's Columbia Presbyterian Medical Center, Ehrhardt cites cases of girls whose adrenal glands, because of an enzyme defect, produced abnormally large amounts of androgens while they were still in the womb. "We find that they are extremely tomboyish," she says. "They are career oriented, and spend little time with dolls. And we've just learned that boys exposed before birth to drugs that contain high doses of feminizing hormones engage in less roughhousing than other boys."

Animal Behavior

11 Ehrhardt admits that labeling the pursuit of a career masculine and playing with dolls feminine seems like stereotyping. To substantiate her evidence, she has compared her results with those obtained

from studies of animals, whose gender behavior is rigid and easily defined.

12 Animal physiologists first made the connection between hormones and behavior in 1849, when the German scientist Arnold Berthold castrated roosters and found that they stopped fighting with other roosters and lost interest in attracting hens. When he transplanted the testicles into the abdominal cavities of the castrated birds, the roosters became aggressive again. Observing that the transplanted testicles did not develop connections with the rooster's nervous system but did develop connections with its circulatory system, he speculated that their influence on behavior came from a blood-borne substance, which was later identified as a hormone.

13 In 1916, Frank Lillie, a Canadian physiologist, noticed that the freemartin, a genetically female (X-X) cow that looks and acts like a male, always had a male twin. He speculated that the freemartin's gonads were masculinized in the womb by hormones secreted by the testicles of the twin.

14 Fascinated by this finding, scientists began using testosterone to make "freemartin" guinea pigs, rats, monkeys, and dogs. This set the stage for the landmark experiment conducted at the University of Kansas in 1959 by physiologists William Young and Robert Goy.

15 "We injected pregnant guinea pigs with huge amounts of testosterone," explains Goy. "This produced a brood of offspring in which those that were genetically female had male genitalia as well as ovaries." When the females were 90 days old, the researchers removed their ovaries and injected some of them with still more testosterone. The injected females began to act like males, mounting other females and trying to dominate the group. Says Goy, "We realized that we had changed the sex of the guinea pig's brain."

16 The researchers concluded that hormones affect behavior in two ways. Before birth, hormones imprint a code on the brain, "just as light can stamp an image on film," Goy says. "Later, throughout life, other hormones activate the code, much as a developer brings out an image on film. Whether the animal behaves like a male or a female depends on the code."

17 Goy has spent the past two decades proving that theory for a whole range of species, including the rhesus monkey. Now at the Primate Research Center at the University of Wisconsin in Madison, he has found that masculinized monkeys display sexual behavior that

ranges from female to male in direct proportion to the amount of testosterone they are given while in the womb and throughout life. "It doesn't much matter whether it's rough-and-tumble play, mounting peers, or attempting to dominate the group," he says. "It's all related to the duration of treatment."

18 Perhaps more important, Goy has found that by varying the treatment he can produce monkeys that are physically female but behave like males. This is proof, he says, "that these animals behave like boys because of masculinizing hormones, not because of a male appearance that causes the other animals to treat them like boys."

19 Like the human brain, the brain of the rhesus monkey has a highly elaborate and convoluted cortex. But Goy believes that monkeys can be compared with people only up to a point. For while primitive drives may be similar, he says, human beings are guided by their culture to a greater degree than monkeys. "Nevertheless," he adds, "there are instances when people seem to be less bound by culture. Then they begin to look very much like our monkeys."

Biochemistry

20 Other scientists have substantiated this evidence with hard biochemical data. To learn where sex hormones operate, neurobiologist Donald Pfaff of New York City's Rockefeller University injected various animals with radioactive hormones and removed their brains. He cut each brain into paper-thin sections, then placed each section on film sensitive to radioactivity. He thus made maps showing that the hormones collected at specific places, now called receptor sites, are similarly located in the brains of species ranging from fish to rats to the rhesus monkey.

21 The primary site for hormone action, Pfaff saw, was the hypothalamus, a primitive structure at the base of the brain stem. That made sense, because the hypothalamus is the center for sex drive and copulatory behavior. "But the most intriguing thing," says Pfaff, "may be the receptors found in the amygdala [a part of the brain above each ear]. During the 1960s, surgeons found that when they destroyed the amygdala, patients with fits of aggression became completely passive. So we now suspect that sex hormones may control aggression, even fear." Neurologist Bruce McEwen, also of Rockefeller, recently found estrogen receptors in the cerebral cortex of the rat—receptors that disappear three weeks after birth. The cortex controls thought and

cognition, but McEwen does not know the significance of these receptors.

22 The receptors are located at the same sites in both sexes, but because each sex has its own characteristic mix of hormones, male and female brains function differently. To unravel the secret of hormone operation, McEwen has been analyzing the chemistry of the rat brain. He has discovered that receptor sites are hormone specific; a testosterone site, for example, is insensitive to estrogen. Perhaps more important, he has learned that once hormones pair up with receptors, they mold the structure of the brain by directing nerve cells to manufacture proteins. Early in life, the proteins build nerve cells, creating permanent structures that may exist in the brain of one sex but not the other. Later in life, the proteins produce the chemicals that enable one nerve cell to communicate with another, and precipitate various kinds of sexual behavior.

23 McEwen and Pfaff have not dissected human brains, but they feel justified in applying some of their findings to people. For, as Pfaff explains, evolution is a conservationist. "As new species evolved, nature didn't throw away old parts of the brain," he says. "Rather, new systems were added. Everyone has a fish brain deep inside. Outside the fish brain there is a reptilian brain, depressingly similar to the way it would look in a lizard. Wrapped around the reptilian brain there is a mammalian brain, and then, finally, the cerebral cortex in such animals as monkeys and human beings." McEwen thinks that the receptors in the hypothalamus probably have similar effects in people and rats. "The difference," he says, "is that human beings can override their primitive drives with nerve impulses from the powerful cerebral cortex."

Anatomy

24 Anatomical evidence that sex hormones change the structure of the brain came recently from Roger Gorski, a neuroendocrinologist at the University of California at Los Angeles. Examining the hypothalamus in rats, he found a large cluster of nerve cells in the males and a small cluster in females. By giving a female testosterone shortly after birth, he created a large cluster of cells in her hypothalamus that resembled that in the male. If he castrated a male after birth, its cell cluster shrank. Gorski has no idea what the cell structure signifies, but he does know that it varies with changes in sexual behavior.

25 The anatomical differences do not stop there. Fernando Nottebohm, of Rockefeller, has discovered a large brain-cell cluster in the male canary and a small one in the female. These cells are not in the spinal cord or the hypothalamus but in the forebrain—the songbird equivalent of the cerebral cortex, the part that controls thought and cognition.

26 The function that Nottebohm studied was song. Only the male songbird can sing, and the more intricate the song the more females he attracts. That takes brainwork, says Nottebohm. "The canary puts songs together just as the artist creates. A large collection of syllables can be combined in infinite ways to form a repertoire in which each story is unique.

27 Until Nottebohm discovered the large cluster of male brain cells that control the muscles of the syrinx, the singing organ, he had assumed that male and female brains were anatomically identical. He found that if he gave female canaries testosterone before they hatched and again during adulthood, they could learn to sing. When he studied the brains of the singing females, he found that their cell clusters had grown. Says Nottebohm, "The intriguing thing is that the size of the repertoire was more or less proportional to the size of the cell clusters."

28 Scientists studying mammals have also discovered anatomical differences between the sexes in the thinking part of the brain—in this case, the cerebral cortex of the rat. Marian Diamond, of the University of California at Berkeley, discovered that in the male rat the right hemisphere of the cortex was thicker than the left—and that in the female the left was thicker than the right. But if she castrated the male rat at birth or removed the ovaries from the female, she could alter the pattern. Administering female hormones to males and male hormones to females also affected the width of the cortex. Says Diamond, "Hormones present during pregnancy, hormones present in the birth-control pill, all affect the dimensions of the cortex."

29 Jerre Levy, a neuropsychologist at the University of Chicago, is encouraged by Diamond's findings because they provide strong anatomical evidence for her theory: the cortex is different in men and women, largely because of hormones that early in life alter the organization of the two hemispheres.

30 Levy is responsible for much of what is known about the human brain's laterality—the separation of the roles performed by the right

and left hemispheres. Levy began her work in this field in the 1960s, when she was studying "split brain" patients, epileptics whose hemispheres had been surgically separated as a means of controlling violent seizures. The researchers found that the hemispheres could operate independently of each other, somewhat like two minds in a single head. The right hemisphere specialized in the perception of spatial relationships, like those in mazes and solid geometry, and the left controlled language and rote memory.

31 Levy has found that these abilities vary with gender. In test after test, men excelled in spatial reasoning and women did better with language. Fascinated by the discrepancy, she decided to test laterality in normal people and based her experiments on a well known fact: light and sound coming from the visual and auditory fields on one side of the head travel to the hemisphere on the other side for processing.

32 She discovered that the information from the right is perceived more acutely by women while information from the left is perceived more acutely by men. She concluded that the right hemisphere dominates the masculine brain, and the left the feminine.

33 Levy points to the work of neuropsychologist Deborah Waber of Harvard Medical School, who found that children reaching puberty earlier than normal have brains that are less lateralized—that is, their left and right hemispheres seem to share more tasks. Because girls generally reach puberty two years before boys, these findings have caused speculation that the bundle of nerve connections, the corpus callosum, between the two hemispheres of the female brain have less time to lateralize, or draw apart, during puberty. If that is true, says Levy, it could help to explain female intuition, as well as male superiority in mechanics and math. The two intimately connected hemispheres of the female brain would communicate more rapidly—an advantage in integrating all the detail and nuance in an intricate situation, but according to Levy a disadvantage "when it comes to homing in on just a few relevant details." With less interference from the left hemisphere, Levy says, a man could "use his right hemisphere more precisely in deciphering a map or finding a three-dimensional object in a two-dimensional representation."

34 All this brings Levy back to hormones. She thinks that the estrogen that changes the size of the cortex in Marian Diamond's rats may also change the size and organization of the human cortex. Her new

tests are designed to study the organization of the cerebral cortex in people with hormonal abnormalities—girls who produce an excess of androgen and boys who are exposed to large amounts of estrogen before birth.

35 Levy has ambitious plans for future research, including scans of living brains and tests of babies whose mothers have undergone stress during pregnancy. Much remains to be done, for though the existence of physical differences between male and female brains now seems beyond dispute, the consequences are unclear. Talent in math, for example, is obviously not confined to men nor talent in languages to women; the subtleties seem infinite. Already the new findings promise to color the modern view of the world. But the implications can easily be misconstrued.

36 Gunther Dörner, an East German hormone researcher, has claimed that he can put an end to male homosexuality by injecting pregnant women with testosterone. Dörner bases his theory on studies done by two American researchers, who subjected pregnant rats to stress by confining them in small cages under bright lights. they found that the rats' male offspring had low levels of testosterone during certain critical periods, and exhibited homosexual behavior. Dörner concluded that stress on pregnant females alters sexual preference patterns in the brains of their male offspring, and that this finding applies to human beings as well. His suggested antidote: testosterone.

37 His conclusions appall the American researchers, who agree that mothers under stress produce male offspring with abnormal behavior, but argue that Dörner has gone too far. Dörner's work is supported by the East German government, which is notorious in its aversion to homosexuality, and American scientists fear that he may get a chance to put his ideas into practice on human beings.

38 Another example of misinterpretation is the article that appeared in *Commentary* magazine in December [1980], citing the "latest" in brain research as an argument against equal rights for women. This angers Anne Petersen. "A lot of people have been making a lot of political hoopla about our work," she says. "They've used it to say that the women's movement will fail, that women are inherently unequal. Our research shows nothing of this sort, of course. There are things that men do better, and things that women do better. It's very important to differentiate between the inferences and the scientific findings."

39 These findings could influence fields ranging from philosophy, psychiatry, and the arts to education, law, and medicine. If women are indeed at a disadvantage in mastering math, there could be different methods of teaching, or acceptance of the fact that math is not important for certain jobs. For example, tests of mathematical competence have been used as criteria for admission to law school, where math is barely used; tests of spatial ability have been used to screen people for all types of nontechnical pursuits. If scientists can prove that such tests discriminate unnecessarily against women, hiring policies could be changed. Eventually, psychiatrists and lawyers may have to assess their male and female clients in a new light. And brain surgeons may have to consider the sex of a patient before operating. For if the two hemispheres of the brain are more intimately connected in women than in men, then women may be able to control a function like speech with either hemisphere. Surgeons could feel confident that a woman would recover the ability to talk, even if her normal speech center were destroyed; they might proceed with an operation that they would hesitate to perform on a man.

40 Investigators have made amazing progress in their work on the sexes and the brain, but they have really just begun. They will have to link hundreds of findings from widely diverse areas of brain science before they can provide a complete explanation for the shared, but different, humanity of men and women.

Suggestions for Discussion and Writing

1. According to Weintraub, there have been many scientific studies that show differences in female and male brains. How persuaded are you by the evidence she presents in support of the "different brains" arguments? What other kinds of evidence might make her essay more convincing than you now think it is? If you have read Bethell's essay on evidence for evolution, does that alter the kinds of evidence you expect to see in an argument like Weintraub's?

2. What kind of tone does Weintraub take in her introduction? What consequences of the argument that she wants to make does she seem to be worrying about?

3. The research that Weintraub cites raises a question of "nature vs. nurture"; in other words, are certain qualities inborn or the products of our society? Which category, nature or nurture, does the research Weintraub summarizes seem to support? How conclusive is this evidence?

4. Whom do you think Weintraub envisioned as the readers of this essay? If you were going to write a story on this topic for a local publication (pick a local newspaper, magazine, or news program), whom do you think your audience would be? Which of Weintraub's strategies for appealing to her audience could you use?

5. This essay is now ten years old. What kind of evidence has arisen in the last decade to further Weintraub's argument? Which "side" is "winning"? Do you think we will ever be able to answer this question conclusively? Why do you think people want a definitive answer to this question?

Barry Lopez

Encounter on the Tundra

From his first publications, BARRY LOPEZ has been respected as an authoritative writer on natural history and environmental issues. Of late, he has used these issues to explore larger moral questions as well. Born in New York in 1945, Lopez was raised in the West and returned there first to attend school and to write. He and his family settled in western Oregon, where he devoted himself to writing full time. He is the author of *Desert Notes* (1976), *Of Wolves and Men* (1978), *River Notes* (1979), *Arctic Dreams* (1986), *Crossing Open Ground* (1988), and *The Rediscovery of North America* (1990).

Lopez has been particularly interested in showing how humanity's view of nature is a mirror image of their understanding of themselves. He told an interviewer, "My interest in a story is to illuminate a set of circumstances that bring some understanding of human nature, enough at least so that a reader can identify with it and draw some vague sense of hope or sustenance or deep feeling and in some way be revived." He told another interviewer, "You believe the story, whether it's nonfiction or fiction, has a power to elevate or to heal or to illuminate, to provide hope, to in some way give an individual life greater dimensions. These are all community obligations." This essay is taken from *Arctic Dreams.*

In certain parts of the Arctic—Lancaster Sound, the shores of Queen Maud Gulf, the Mackenzie River Delta, northern Bering Sea, the Yukon-Kuskokwin Delta—great concentrations of wildlife seem to belie violent fluctuations in this ecosystem. The Arctic seems resplendent with life. But these are summer concentrations, at well-known oases, widely separated over the land; and they consist largely of migratory creatures—geese, alcids, and marine mammals. When the rivers and seas freeze over in September they will all be gone The winter visitor will find only caribou and muskoxen, and occasionally arctic hares, concentrated in any number, and again only in a few places.

2 All life, of course, cannot fly or swim or walk away to a warmer climate. When winter arrives, these animals must disperse to areas where they will have a good chance to find food and where there is some protection from the weather. A few hibernate for seven or eight months. Voles and lemmings go to ground too, but remain active all winter. Wolves shift their home ranges to places where caribou and moose are concentrated. Arctic foxes follow polar bears out onto the sea ice, where they scavenge the bear's winter kills. Arctic hares seek out windblown slopes where vegetation is exposed. All these resident animals have a measure of endurance about them. They expect to see you, as unlikely as it may seem, in the spring.

3 In my seasonal travels the collared lemming became prominent in my mind as a creature representative of winter endurance and resiliency. When you encounter it on the summer tundra, harvesting lichen or the roots of cotton grass, it rises on its back feet and strikes a posture of hostile alertness that urges you not to trifle. Its small size is not compromising; it displays a quality of heart, all the more striking in the spare terrain.

4 Lemmings are ordinarily sedentary, year-round residents of lo-

cal tundra communities. They came into the central Arctic at the end of the Pleistocene some 8,000 years ago, crossing great stretches of open water and extensive rubble fields of barren sea ice to reach the places they live in today. In winter lemmings live under an insulating blanket of snow in a subnivean landscape, a dark, cool, humid world of quiet tunnels and windless corridors. They emerge in spring to a much brighter, warmer, and infinitely more open landscape—where they are spotted by hungry snowy owls and parasitic jaegers and are hunted adroitly by foxes and short-tailed weasels. In most years, in most places, there is not much perplexing about this single link in several arctic food chains. In some places, every three or four years, however, the lemming population explodes. Lemmings emerge from their subnivean haunts in extraordinary numbers and strike out—blindly is the guess—across the tundra.

5 The periodic boom in lemming populations—there are comparable, though more vaguely defined, cycles affecting the periodic rise and fall of snowshoe hare and lynx populations, and caribou and wolf populations—is apparently connected with the failure of the lemmings' food base. The supply of available forage reaches a peak and then collapses, and the lemmings move off smartly in all directions as soon as traveling conditions permit in the spring. Occasionally many thousands of them reach sea cliffs or a swift-moving river; those pushing in the rear force the vanguard into the water to perish.

6 Arctic scientist Laurence Irving, camped once on a gravel bar off the Alaska coast, wrote: "In the spring of a year of climaxing abundance, a lively and pugnacious lemming came into my camp. . . [more] tracks and a dead lemming were seen on the ice several kilometers from shore. The seaward direction of this mad movement was pointless, but it illustrates stamina that could lead to a far dispersal." Irving's regard, of course, is a regard for the animal itself, not for the abstract mechanisms of population biology of which it seems to merely be a part. Its apparently simple life on the tundra suggests it can be grasped, while its frantic migrations make it seem foolish. In the end, it is complex in its behavior, intricately fitted into its world, and mysterious.

7 Whenever I met a collared lemming on a summer day and took its stare I would think: Here is a tough animal. Here is a valuable life. In a heedless moment, years from now, will I remember more machinery here than mind? If it could tell me of its will to survive,

would I think of biochemistry, or would I think of the analogous human desire? If it could speak of the time since the retreat of the ice, would I have the patience to listen?

8 One time I fell asleep on the tundra, a few miles from our camp. I was drowsy with sun and the weight of languid air. I nestled in the tussock heath, in the warm envelope of my down parka; and was asleep in a few moments. When I awoke I did not rise, but slowly craned my head around to see what was going on. At a distance I saw a ground squirrel crouched behind a limestone slab that rose six or eight inches out of the ground like a wall. From its attitude I thought it was listening, confirming the presence of some threat on the other side of the rock, in a shallow draw. After a while it put its paws delicately to the stone and slowly rose up to peer over, breaking the outline of the rock with the crown of its head. Then, with its paws still flat at the rim, it lowered itself and rested its forehead on the rock between its forelegs. The feeling that it was waiting for something deadly to go away was even stronger. I thought: Well, there is a fox over there, or a wolverine. Maybe a bear. He'd better be careful.

9 I continued to stare at him from the warm crevice in the earth that concealed me. If it is a bear, I thought, I should be careful too, not move from here until the ground squirrel loses that tension in its body. Until it relaxes, and walks away.

10 I lay there knowing something eerie ties us to the world of animals. Sometimes the animals pull you backward into it. You share, hunger and fear with them like salt in blood.

11 The ground squirrel left. I went over to the draw beyond the rock but could find no tracks. No sign. I went back to camp mulling the arrangements animals manage in space and in time—their migrations, their patience, their lairs. Did they have intentions as well as courage and caution?

12 Few things provoke like the presence of wild animals. They pull at us like tidal currents with questions of volition, of ethical involvement, of ancestry.

13 For some reason I brooded often about animal behavior and the threads of evolution in the Arctic. I do not know whether it was the reserves of space, the simplicity of the region's biology, its short biological history, striking encounters with lone animals, or the realization of my own capacity to annihilate life here. I wondered where the animals had come from; and where we had come from; and where

each of us was going. The ecosystem itself is only 10,000 years old, the time since the retreat of the Wisconsin ice. The fact that it is the youngest ecosystem on earth gives it a certain freshness and urgency. (Curiously, historians refer to these same ten millennia as the time of civilized man, from his humble beginnings in northern Mesopotamia to the present. Arctic ecosystems and civilized man belong, therefore, to the same, short epoch, the Holocene. Mankind is, in fact, even older than the Arctic, if you consider his history to have begun with the emergence of Cro-Magnon people in Europe 40,000 years ago.)

14 Human beings dwell in the same biological systems that contain the other creatures but, to put the thought bluntly, they are not governed by the same laws of evolution. With the development of various technologies—hunting weapons, protective clothing, fire-making tools; and then agriculture and herding—mankind has not only been able to take over the specific niches of other animals but has been able to move into regions that were formerly unavailable to him. The animals he found already occupying niches in these other areas he, again, either displaced or eliminated. The other creatures have had no choice. They are confined to certain niches—places of food (stored solar energy), water, and shelter—which they cannot leave without either speciating or developing tools. To finish the thought, the same technological advances and the enormous increase in his food base have largely exempted man from the effect of natural controls on the size of his population. Outside of some virulent disease, another ice age, or his own weapons technology, the only thing that promises to stem the continued increase in his population and the expansion of his food base (which now includes oil, exotic minerals, fossil ground water, huge tracts of forest, and so on, and entails the continuing, concomitant loss of species) is human wisdom.

15 Walking across the tundra, meeting the stare of a lemming, or coming on the tracks of a wolverine, it would be the frailty of our wisdom that would confound me. The pattern of our exploitation of the Arctic, our increasing utilization of its natural resources, our very desire to "put it to use," is clear. What is it that is missing, or tentative, in us, I would wonder, to make me so uncomfortable walking out here in a region of chirping birds, distant caribou, and redoubtable lemmings? It is restraint.

16 Because mankind can circumvent evolutionary law, it is incum-

bent upon him, say evolutionary biologists, to develop another law to abide by if he wishes to survive, to not outstrip his food base. He must learn restraint. He must derive some other, wiser way of behaving toward the land. He must be more attentive to the biological imperatives of the system of sun-driven protoplasm upon which he, too, is still dependent. Not because he must, because he lacks inventiveness, but because herein is the accomplishment of the wisdom that for centuries he has aspired to. Having taken on his own destiny, he must now think with critical intelligence about where to defer.

17 A Yup'ik hunter on Saint Lawrence Island once told me that what traditional Eskimos fear most about us is the extent of our power to alter the land, the scale of that power, and the fact that we can easily effect some of these changes electronically, from a distant city. Eskimos, who sometimes see themselves as still not quite separate from the animal world, regard us as a kind of people whose separation may have become too complete. They call us, with a mixture of incredulity and apprehension, "the people who change nature."

Suggestions for Discussion and Writing

1. What is your picture of the Arctic tundra? Do you expect it to be as full of life as Lopez pictures it to be? Why does he begin the essay with this picture of natural diversity?

2. What is a lemming? What behaviors are most associated with them? Why does Lopez include a long discussion of lemmings in this essay?

3. Lopez says that human beings "are not governed by the same laws of evolution" as the rest of nature. What are the consequences of the difference he sees? Do Bethell's and Eldredge's arguments help you understand what Lopez means?

4. Why does Lopez feel that humans must learn restraint? What kinds of restraint is he talking about? Of the other authors you have read, who would most agree with him? Who would disagree the most?

5. Today we hear calls from many places for humanity to respect the environment (for instance, saving whales, the rainforest, the condor). Assume you are writing the cover story for a magazine like *Time* or *Newsweek*. Pick one of these calls that interests you and write a report on the kinds of nature being threatened, and how people are responding to this threat.

Loren Eiseley

Science and the Sense of the Holy

LOREN EISELEY, born in Lincoln, Nebraska, in 1907, received his Ph.D. from the University of Pennsylvania during the Depression, and taught at the University of Kansas and at Oberlin before returning to Penn at the end of the Second World War. He spent the rest of his distinguished writing and teaching career as Curator of Early Man at the University of Pennsylvania Museum until his death in 1977.

Whether writing about a fossil, a pile of bones, or a long-dead scientist, Eiseley infused their lives with beauty and significance. Some "pure" scientists criticized him for this "poeticizing" and for the occasional difficulty of his prose, but writers as diverse as W.H. Auden and Ray Bradbury defended him. Eiseley believed strongly that people must develop their abilities to imagine and to dream; he feared that a world based only on reason and the rational would destroy humanity. In a 1970 letter he wrote, "What I really fear is that man will ruin the planet before he departs. I have sometimes thought. . . what a beautiful ruin (Manhattan) would make in heaps of fallen masonry, with the forest coming back. Now I fear for the forest itself." This is an essay from *The Star Thrower* (1978) .

I

When I was a young man engaged in fossil hunting in the Nebraska badlands I was frequently reminded that the ravines, washes, and gullies over which we wandered resembled the fissures in a giant exposed brain. The human brain contains the fossil memories of its past—buried but not extinguished moments—just as this more formidable replica contained deep in its inner stratigraphic convolutions earth's past in the shape of horned titanotheres and stalking, dirk-toothed cats. Man's memory erodes away in the short space of a lifetime. Jutting from the coils of the earth brain over which I clambered were the buried remnants, the changing history, of the entire age of mammals—millions of years of vanished daylight with their accompanying traces of volcanic outbursts and upheavals. It may well be asked why this analogy of earth's memory should so preoccupy the mind of a scientist as to have affected his entire outlook upon nature and upon his kinship with—even his concern for—the plant and animal world about him.

2 Perhaps the problem can best be formulated by pointing out that there are two extreme approaches to the interpretation of the living world. One was expressed by Charles Darwin at the age of twenty-eight; one by Sigmund Freud in his mature years. Other men of science have been arrayed on opposite sides of the question, but the eminence of these two scholars will serve to point up a controversy that has been going on since science arose, sometimes quietly, sometimes marked by vitriolic behavior, as when a certain specialist wedded to his own view of the universe hurled his opponent's book into his wastebasket only to have it retrieved and cherished by a graduate student who became a lifelong advocate of the opinions reviled by his mentor. Thus it is evident that, in the supposed objective world of science, emotion and temperament may play a role in our selection of the mental tools with which we choose to investigate nature.

3 Charles Darwin, at a time when the majority of learned men looked upon animals as either automatons or creatures created merely for human exploitation, jotted thoughtfully into one of his early journals upon evolution the following observation:

4 "If we choose to let conjecture run wild, then animals, our fellow brethren in pain, disease, suffering and famine—our slaves in the most laborious works, our companions in our amusements—they may partake of our origin in one common ancestor—we may be all netted together."

5 What, we may now inquire, is the world view here implied, one way in which a great scientist looked upon the subject matter that was

to preoccupy his entire working life? In spite of the fact that Darwin was, in his later years, an agnostic, in spite of confessing he was "in thick mud" so far as metaphysics was concerned, the remark I have quoted gives every sign of that feeling of awe, of dread of the holy playing upon nature, which characterizes the work of a number of naturalists and physicists down even to the present day. Darwin's remark reveals an intuitive sensitivity to the life of other creatures about him, an attitude quite distinct from that of the laboratory experimentalist who is hardened to the infliction of pain. In addition, Darwin's final comment that we may be all netted together in one gigantic mode of experience, that we are in a mystic sense one single diffuse animal, subject to joy and suffering beyond what we endure as individuals, reveals a youth drawn to the world of nature by far more than just the curiosity to be readily satisfied by the knife or the scalpel.

6 If we turn to Sigmund Freud by way of contrast we find an oddly inhibited reaction. Freud, though obviously influenced by the elegant medical experimenters of his college days, groped his way alone, and by methods not subject to quantification or absolute verification, into the dark realms of the subconscious. His reaction to the natural world, or at least his feelings and intuitions about it, are basically cold, clinical, and reserved. He of all men recognized what one poet has termed "the terrible archaeology of the brain." Freud states that "nothing once constructed has perished, and all the earlier stages of development have survived alongside the latest." But for Freud, convinced that childhood made the man, adult reactions were apt to fall under the suspicion of being childhood ghosts raised up in a disguised fashion. Thus, insightful though he could be, the very nature of his study of man tended to generate distrust of that outgoing empathy we observed in the young Darwin. "I find it very difficult to work with these intangible qualities," confessed Freud. He was suspicious of their representing some lingering monster of childhood, even if reduced in size. Since Freud regarded any type of religious feeling—even the illuminative quality of the universe—as an illusion, feelings of awe before natural phenomena such as that manifested by Darwin were to him basically remnants of childhood and to be dismissed accordingly.

7 In *Civilization and Its Discontents* Freud speaks with slight condescension of a friend who claimed a sensation of eternity, something limitless, unbounded—"oceanic," to use the friend's expression.

The feeling had no sectarian origin, no assurance of immortality, but implied just such a sense of awe as might lie at the root of the religious impulse. "I cannot," maintained Freud, "discover this 'oceanic' impulse in myself." Instead he promptly psychoanalyzes the feeling of oneness with the universe in the child's pleasure ego which holds to itself all that is comforting; in short, the original ego, the infant's ego, included everything. Later, by experience, contended Freud, our adult ego becomes only a shrunken vestige of that far more extensive feeling which "expressed an inseparable connection. . . with the external world."

8 In essence, then, Freud is explaining away one of the great feelings characteristic of the best in man by relegating it to a childhood atavistic survival in adult life. The most highly developed animals, he observes, have arisen from the lowest. Although the great saurians are gone, the dwarfed crocodile remains. Presumably if Freud had completed the analogy he would have been forced to say that crocodilian adults without awe and with egos shrunken safely into their petty concerns represented a higher, more practical evolutionary level than the aberrant adult who persists in feelings of wonder before which Freud recoiled with a nineteenth-century mechanist's distaste, although not without acknowledging that this lurking childlike corruption might be widespread. He chose to regard it, however, as just another manifestation of the irrational aspect of man's divided psyche.

9 Over six decades before the present, a German theologian, Rudolf Otto, had chosen for his examination what he termed *The Idea of the Holy (Das Heilige).* Appearing in 1917 in a time of bitterness and disillusionment, his book was and is still widely read. It cut across denominational divisions and spoke to all those concerned with that *mysterium tremendum,* that very awe before the universe which Freud had sighed over and dismissed as irrational. I think it safe to affirm that Freud left adult man somewhat shrunken and misjudged—misjudged because some of the world's scientists and artists have been deeply affected by the great mystery, less so the child at one's knee who frequently has to be disciplined to what in India has been called the "opening of the heavenly eye."

10 Ever since man first painted animals in the dark of caves he has been responding to the holy, to the numinous, to the mystery of being and becoming, to what Goethe very aptly called "the weird portentous." Something inexpressible was felt to lie behind nature. The

bear cult, circumpolar in distribution and known archaeologically to extend into Neanderthal times, is a further and most ancient example. The widespread beliefs in descent from a totemic animal, guardian helpers in the shapes of animals, the concept of the game lords who released or held back game to man are all part of a variety of a sanctified, reverent experience that extends from the beautiful rock paintings of South Africa to the men of the Labradorean forests or the Plains Indian seeking by starvation and isolation to bring the sacred spirits to his assistance. All this is part of the human inheritance, the wonder of the world, and nowhere does that wonder press closer to us than in the guise of animals which, whether supernaturally as in the caves of our origins or, as in Darwin's sudden illumination, perceived to be, at heart, one form, one awe-inspiring mystery, seemingly diverse and apart but derived from the same genetic source. Thus the *mysterium* arose not by primitive campfires alone. Skins may still prickle in a modern classroom.

11 In the end, science as we know it has two basic types of practitioners. One is the educated man who still has a controlled sense of wonder before the universal mystery, whether it hides in a snail's eye or within the light that impinges on that delicate organ. The second kind of observer is the extreme reductionist who is so busy stripping things apart that the tremendous mystery has been reduced to a trifle, to intangibles not worth troubling one's head about. The world of the secondary qualities—color, sound, thought—is reduced to illusion. The *only* true reality becomes the chill void of ever-streaming particles.

12 If one is a biologist this approach can result in behavior so remarkably cruel that it ceases to be objective but rather suggests a deep grain of sadism that is not science. To list but one example, a recent newspaper article reported that a great urban museum of national reputation had spent over a half-million dollars on mutilating experiments with cats. The experiments are too revolting to chronicle here and the museum has not seen fit to enlighten the public on the knowledge gained at so frightful a cost in pain. The cost, it would appear, lies not alone in animal suffering but in the dehumanization of those willing to engage in such blind, random cruelty. The practice was defended by museum officials, who in a muted show of scientific defense maintained the right to study what they chose "without regard to its demonstrable practical value."

13 This is a scientific precept hard to override since the days of Galileo, as the official well knew. Nevertheless, behind its seamless façade of probity many terrible things are and will be done. Blaise Pascal, as far back as the seventeenth century, foresaw our two opposed methods. Of them he said: "There are two equally dangerous extremes, to shut reason out, and to let nothing else in." It is the reductionist who, too frequently, would claim that the end justifies the means, who would assert reason as his defense and let that *mysterium* which guards man's moral nature fall away in indifference, a phantom without reality.

14 "The whole of existence frightens me," protested the philosopher Søren Kierkegaard; "from the smallest fly to the mystery of the Incarnation, everything is unintelligible to me, most of all myself." By contrast, the evolutionary reductionist Ernst Haeckel, writing in 1877, commented that "the cell consists of matter. . . composed chiefly of carbon with an admixture of hydrogen, nitrogen and sulphur. These component parts, properly united, produce the soul and body of the animated world, and suitably nourished become man. With this single argument the mystery of the universe is explained, the Deity annulled and a new era of infinite knowledge ushered in." Since these remarks of Haeckel's, uttered a hundred years ago, the genetic alphabet has scarcely substantiated in its essential intricacy Haeckel's carefree dismissal of the complexity of life. If anything, it has given weight to Kierkegaard's wary statement or at least heightened the compassionate wonder with which we are led to look upon our kind.

15 "A conviction akin to religious feeling of the rationality or intelligibility of the world lies behind all scientific work of a high order," says Albert Einstein. Here once more the eternal dichotomy manifests itself. Thoreau, the man of literature, writes compassionately, "Shall I not have intelligence with the earth? Am I not partly leaves and vegetable mould myself?" Or Walt Whitman, the poet, protests in his *Song of Myself:* "whoever walks a furlong without sympathy walks to his own funeral drest in a shroud."

> "Magnifying and applying come I"—he thunders—
> "Outbidding at the start the old cautious hucksters. . .
> Not objecting to special revelations, considering a curl of smoke or a hair on the back of my hand just as curious as any revelation."

16 Strange, is it not, that so many of these voices are not those of children, but those of great men—Newton playing on the vast shores of the universe, or Whitman touched with pity or Darwin infused with wonder over the clambering tree of life. Strange, that all these many voices should be dismissed as the atavistic yearnings of an unreduced childlike ego seeking in "oceanic" fashion to absorb its entire surroundings, as though in revolt against the counting house, the laboratory, and the computer.

II

17 Not long ago in a Manhattan art gallery there were exhibited paintings by Irwin Fleminger, a modernist whose vast lawless Martianlike landscapes contain cryptic human artifacts. One of these paintings attracted my eye by its title: "Laws of Nature." Here in a jumbled desert waste without visible life two thin laths had been erected a little distance apart. Strung across the top of the laths was an insubstantial string with even more insubstantial filaments depending from the connecting cord. The effect was terrifying. In the huge inhuman universe that constituted the background, man, who was even more diminished by his absence, had attempted to delineate and bring under natural law an area too big for his comprehension. His effort, his "law," whatever it was, denoted a tiny measure in the midst of an ominous landscape looming away to the horizon. The frail slats and dangling string would not have sufficed to fence a chicken run.

18 The message grew as one looked. With all the great powers of the human intellect we were safe, we understood, in degree, a space between some slats and string, a little gate into the world of infinitude. The effect was crushing and it brought before one that sense of the "other" of which Rudolf Otto spoke, the sense beyond our senses, unspoken awe, or, as the reductionist would have it, nothing but waste. There the slats stood and the string drooped hopelessly. It was the natural law imposed by man, but outside its compass, again to use the words of Thoreau, was something terrific, not bound to be kind to man. Not man's at all really—a star's substance totally indifferent to life or what laws life might concoct. No man would greatly extend that trifling toy. The line sagged hopelessly. Man's attempt had failed, leaving but an artifact in the wilderness. Perhaps, I thought, this is man's own measure. Perhaps he has already gone. The crepitation at my

spine increased. I felt the mood of the paleolithic artists, lost in the mysteries of birth and coming, as they carved pregnant beasts in the dark of caves and tried by crayons to secure the food necessarily wrung from similar vast landscapes. Their art had the same holy quality that shows in the ivory figurines, the worship before the sacred mother who brought man mysteriously into the limited world of the cave mouth.

19 The numinous then is touched with superstition, the reductionist would say, but all the rituals suggest even toward hunted animals a respect and sympathy leading to ceremonial treatment of hunted souls; whereas by contrast in the modern world the degradation of animals in experiments of little, or vile, meaning, were easily turned to the experimental human torture practiced at Dachau and Buchenwald by men dignified with medical degrees. So the extremes of temperament stand today: the man with reverence and compassion in his heart whose eye ranges farther than the two slats in the wilderness, and the modern vandal totally lacking in empathy for life beyond his own, his sense of wonder reduced to a crushing series of gears and quantitative formula, the educated vandal without mercy or tolerance, the collecting man that I once tried to prevent from killing an endangered falcon, who raised his rifle, fired, and laughed as the bird tumbled at my feet. I suppose Freud might have argued that this was a man of normal ego, but I, extending my childlike mind into the composite life of the world, bled accordingly.

20 Perhaps Freud was right, but let us look once more at this brain that in many distinguished minds has agonized over life and the mysterious road by which it has come. Certainly, as Darwin recognized, it was not the tough-minded, logical inductionists of the early nineteenth century who in a deliberate distortion of Baconian philosophy solved the problem of evolution. Rather, it was what Darwin chose to call "speculative" men, men, in other words, with just a touch of the numinous in their eye, a sense of marvel, a glimpse of what was happening behind the visible, who saw the whole of the living world as though turning in a child's kaleidoscope.

21 Among the purely human marvels of the world is the way the human brain after birth, when its cranial capacity is scarcely larger than that of a gorilla or other big anthropoid, spurts onward to treble its size during the first year of life. The human infant's skull will

soar to a cranial capacity of 950 cubic centimeters while the gorilla has reached only 380 cubic centimeters. In other words, the human brain grows at an exponential rate, a spurt which carries it almost to adult capacity at the age of fourteen.

22 This clever and specifically human adaptation enables the human offspring successfully to pass the birth canal like a reasonably small-headed animal, but in a more larval and helpless condition than its giant relatives. The brain burgeons after birth, not before, and it is this fact which enables the child, with proper care, to assimilate all that larger world which will be forever denied to its relative the gorilla. The big anthropoids enjoy no such expansion. Their brains grow without exponential quickening into maturity. Somewhere in the far past of man something strange happened in his evolutionary development. His skill has enhanced its youthful globularity; he has lost most of his body hair and what remains grows strangely. He demands, because of his immature emergence into the world, a lengthened and protected childhood. Without prolonged familial attendance he would not survive, yet in him reposes the capacity for great art, inventiveness, and his first mental tool, speech, which creates his humanity. He is without doubt the oddest and most unusual evolutionary product that this planet has yet seen.

23 The term applied to this condition is neoteny, or pedomorphism. Basically the evolutionary forces, and here "forces" stands for complete ignorance, seem to have taken a roughhewn ordinary primate and softened and eliminated the adult state in order to allow for a fantastic leap in brain growth. In fact, there is a growing suspicion that some, at least, of the African fossils found and ascribed to the direct line of human ascent in eastern Africa may never, except for bipedalism and some incipient tool-using capacities, have taken the human road at all.

24 Some with brains that seem to have remained at the same level through long ages have what amounts quantitatively to no more than an anthropoid brain. Allowing for upright posture and free use of the hand, there is no assurance that they spoke or in any effective way were more than well-adapted bipedal apes. Collateral relatives, in short, but scarcely to be termed men. All this is the more remarkable because their history can now be traced over roughly five if not six million years—a singularly unprogressive period for a creature destined later to break upon the world with such remarkable results after so long a

period of gestation.

25 Has something about our calculations gone wrong? Are we studying, however necessarily, some bipedal cousins but not ancestral men? The human phylogeny which we seemed well on the way to arranging satisfactorily is now confused by a multiplicity of material contended over by an almost equal number of scholars. Just as a superfluity of flying particles is embarrassing the physicist, so man's evolution, once thought to be so clearly delineated, is showing signs of similar strain. A skull from Lake Rudolf with an estimated capacity of 775 cubic centimeters or even 800 and an antiquity ranging into the three-million-year range is at the human Rubicon, yet much younger fossils are nowhere out of the anthropoid range.

26 Are these all parts of a single variable subhumanity from which we arose, or are some parts of this assemblage neotenous of brain and others not? The scientific exchanges are as stiff with politeness as exchanges between enemies on the floor of the Senate. "Professor so-and-so forgets the difficult task of restoring to its proper position a frontal bone trampled by cattle." A million years may be covertly jumped because there is nothing to be found in it. We must never lose sight of one fact, however: it is by neotenous brain growth that we came to be men, and certain of the South African hominids to which we have given such careful attention seem to have been remarkably slow in revealing such development. Some of them, in fact, during more years than present mankind has been alive seem to have flourished quite well as simple grassland apes.

27 Why indeed should they all have become men? Because they occupied the same ecological niche, contend those who would lump this variable assemblage. But surely paleontology does not always so bind its deliberations. We are here dealing with a gleam, a whisper, a thing of awe in the mind itself, that oceanic feeling which even the hardheaded Freud did not deny existed though he tried to assign it to childhood.

28 With animals whose precise environment through time may overlap, extinction may result among contending forms; it can and did happen among men. But with the first stirrings of the neotenous brain and its superinduced transformation of the family system a new type of ecological niche had incipiently appeared—a speaking niche, a wondering niche which need not have been first manifested in tools but in family organization, in wonder over what lay over the next hill or

what became of the dead. Whether man preferred seeds or flesh, how he regarded his silent collateral relatives, may not at first have induced great competition. Only those gifted with the pedomorphic brain would in some degree have fallen out of competition with the real. It would have been their danger and at the same time their beginning triumph. They were starting to occupy, not a niche in nature, but an invisible niche carved into thought which in time would bring them suffering, superstition, and great power.

29 It cannot, in the beginning, be recognized clearly because it is not a matter of molar teeth and seeds, or killer instincts and ill-interpreted pebbles. Rather it was something happening in the brain, some blinding, irradiating thing. Until the quantity of that gray matter reached the threshold of human proportions no one could be sure whether the creature saw with a human eye or looked upon life with even the faint stirrings of some kind of religious compassion.

30 The new niche in its beginnings is invisible; it has to be inferred. It does not lie waiting to be discovered in a pebble or a massive molar. These things are important in the human dawn but so is the mystery that ordained that mind should pass the channel of birth and then grow like a fungus in the night—grow and convolute and overlap its older buried strata, while a 600-pound gorilla retains by contrast the cranial content of a very small child. When man cast off his fur and placed his trust in that remarkable brain linked by neural pathways to his tongue he had potentially abandoned niches for dreams. Henceforth the world was man's niche. All else would live by his toleration—even the earth from which he sprang. Perhaps this is the hardest, most expensive lesson the layers of the fungus brain have yet to learn: that man is not as other creatures and that without the sense of the holy, without compassion, his brain can become a gray stalking horror—the deviser of Belsen.

31 Its beginning is not the only curious thing about that brain. There are some finds in South Africa dating into immediately postglacial times that reveal a face and calvaria more "modern" in appearance, more pedomorphic, than that of the average European. The skull is marked by cranial capacities in excess of 1700 cubic centimeters—big brained by any standards. The mastoids are childlike, the teeth reduced, the cranial base foreshortened. These people, variously termed Boskopoid or Strandlooper, have, in the words of one anthropologist, "the amazing cranium to face ratio of almost five to one. In Europeans

it is about three to one. Face size has been modernized and subordinated to brain growth." In a culture still relying on coarse fare and primitive implements, the face and brain had been subtly altered in the direction science fiction writers like to imagine as the direction in which mankind is progressing. Yet here the curious foetalization of the human body seems to have outrun man's cultural status, though in the process giving warning that man's brain could still pass the straitened threshold of its birth.

32 How did these people look upon the primitive world into which they found themselves precipitated? History gives back no answer save that here there flourished striking three-dimensional art—the art of the brother animal seen in beauty. Childlike, Freud might have muttered, with childlike dreams, rushed into conflict with the strong, the adult and shrunken ego, the ego that gets what it wants. Yet how strangely we linger over this lost episode of the human story, its pathos, its possible meaning. From whence did these people come? We are not sure. We are not even sure that they derive from one of the groups among the ruck of bipedal wandering apes long ago in Kenya that reveal some relationship to ourselves. Their development was slow, if indeed some of them took that road, the strange road to the foetalized brain that was to carry man outside of the little niche that fed him his tuberous, sandy diet.

33 We thought we were on the verge of solving the human story, but now we hold in our hands gross jaws and delicate, and are unsure of our direction save that the trail is longer than we had imagined twenty years ago. Yet still the question haunts us, the numinous, the holy in man's mind. Early man laid gifts beside the dead, but then in the modern unbelieving world, Ernst Haeckel's world, a renowned philosopher says, "The whole of existence frightens me," or another humbler thinker says, "In the world there is nothing to explain the world" but remembers the gold eyes of the falcon thrown brutally at his feet. He shivers while Freud says, "As for me I have never had such feelings." They are a part of childhood, Freud argues, though there are some who counter that in childhood—yes, even Freud might grant it—the man is made, the awe persists or is turned off by blows or the dullness of unthinking parents. One can only assert that in science, as in religion, when one has destroyed human wonder and compassion, one has killed man, even if the man in question continues to go about his laboratory tasks.

III

34 Perhaps there is one great book out of all American literature which best expresses the clash between the man who has genuine perception and the one who pursues nature as ruthlessly as a hunted animal. I refer to *Moby Dick,* whose narrator, Ishmael, is the namesake of a Biblical wanderer. Every literate person knows the story of Moby Dick and his pursuit by the crazed Captain Ahab who had yielded a leg to the great albino whale. It is the whale and Ahab who dominate the story. What does the whale represent? A symbol of evil, as some critics have contended? Fate, destiny, the universe personified, as other scholars have protested?

35 Moby Dick is "all a magnet," remarks Ahab cryptically at one moment. "And be he agent or be he principal I will wreak my hate upon him." Here, reduced to the deck of a whaler out of Nantucket, the old immortal questions resound, the questions labeled science in our era. Nothing is to go unchallenged. Thrice, by different vessels, Ahab is warned away from his contemplated conquest. The whale does not pursue Ahab, Ahab pursues the whale. If there is evil represented in the white whale it cannot be personalized. The evils of self-murder, of megalomania, are at work in a single soul calling up its foreordained destruction. Ahab heartlessly brushes aside the supplications of a brother captain to aid in the search for his son, lost somewhere in a boat in the trail of the white whale's passing. Such a search would only impede the headlong fury of the pursuit.

36 In Ahab's anxiety to "strike through the mask," to confront "the principal," whether god or destiny, he is denuding himself of all humanity. He has forgotten his owners, his responsibility to his crew. His single obsession, the hidden obsession that lies at the root of much Faustian overdrive in science, totally possesses him. Like Faust he must know, if the knowing kills him, if naught lies beyond. "All my means are sane," he writes, like Haeckel and many another since. "My motive and my object mad."

37 So it must have been in the laboratories of the atom breakers in their first heady days of success. Yet again on the third day Starbuck, the doomed mate, cries out to Ahab, "Desist. See. Moby Dick seeks thee not. It is thou, thou, that madly seekest him." This then is not the pursuit of evil. It is man in his pride that the almighty gods will challenge soon enough. Not for nothing is Moby Dick a white snow hill

rushing through Pacific nights. He carries upon his brow the inscrutability of fate. Agent or principal, Moby Dick presents to Ahab the mystery he may confront but never conquer. There is no harpoon tempered that will strike successfully the heart of the great enigma.

38 So much for the seeking peg-legged man without heart. We know he launched his boats and struck his blows and in the fury of returning vengeance lost his ship, his comrades, and his own life. If, indeed, he pierced momentarily the mask of the "agent," it was not long enough to tell the tale. But what of the sometimes silent narrator, the man who begins the book with the nonchalant announcement, "Call me Ishmael," the man whose Biblical namesake had every man's hand lifted against him? What did he tell? After all, Moby Dick is his book.

39 Ishmael, in contrast to Ahab, is the wondering man, the acceptor of all races and their gods. In contrast to the obsessed Ahab he paints a magnificent picture of the peace that reigned in the giant whale schools of the 1840s, the snuffling and grunting about the boats like dogs, loving and being loved, huge mothers gazing in bliss upon their offspring. After hours of staring in those peaceful depths, "Deep down," says Ishmael, "there I still bathe in eternal mildness of joy." The weird, the holy, hangs undisturbed over the whales' huge cradle. Ishmael knows it, others do not.

40 At the end, when Ahab has done his worst and the *Pequod* with the wounded whale is dragged into the depths amidst shrieking seafowl, it is Ishmael, buoyed up on the calked coffin of his cannibal friend Queequeg, who survives to tell the tale. Like Whitman, like W. H. Hudson, like Thoreau, Ishmael, the wanderer, has noted more of nature and his fellow men than has the headstrong pursuer of the white whale, whether "agent" or "principal," within the universe. The tale is not of science, but it symbolizes on a gigantic canvas the struggle between two ways of looking at the universe: the magnification by the poet's mind attempting to see all, while disturbing as little as possible, as opposed to the plunging fury of Ahab with his cry, "Strike, strike through the mask, whatever it may cost in lives and suffering." Within our generation we have seen the one view plead for endangered species and reject the despoliation of the earth; the other has left us lingering in the shadow of atomic disaster. Actually, the division is not so abrupt as this would imply, but we are conscious as never before in history that there is an invisible line of demarcation, an ethic that

science must sooner or later devise for itself if mankind is to survive. Herman Melville glimpsed in his huge mythology of the white beast that was nature's agent something that only the twentieth century can fully grasp.

41 It may be that those childlike big-brained skulls from Africa are not of the past but of the future, man, not, in Freud's words, retaining an atavistic child's ego, but pushing onward in an evolutionary attempt to become truly at peace with the universe, to know and enjoy the sperm-whale nursery as did Ishmael, to paint in three dimensions the beauty of the world while not to harm it.

42 Yesterday, wandering along a railroad spur line, I glimpsed a surprising sight. All summer long, nourished by a few clods of earth on a boxcar roof, a sunflower had been growing. At last, the car had been remembered. A train was being made up. The box car with its swaying rooftop inhabitant was coupled in. The engine tooted and slowly, with nodding dignity, my plant began to travel.

43 Throughout the summer I had watched it grow but never troubled it. Now it lingered and bowed a trifle toward me as the winds began to touch it. A light not quite the sunlight of this earth was touching the flower, or perhaps it was the watering of my aging eye—who knows? The plant would not long survive its journey but the flower seeds were autumn-brown. At every jolt for miles they would drop along the embankment. They were travelers—travelers like Ishmael and myself, outlasting all fierce pursuits and destined to reemerge into future autumns. Like Ishmael, I thought, they will speak with the voice of the one true agent. "I only am escaped to tell thee."

Suggestions for Discussion and Writing

1. What does the word *holy* bring to mind for you? Is it the kind of word you normally associate with science? Why does Eiseley want you to make this association?

2. What two views of science is Eiseley contrasting in the views of Darwin and Freud? Why does he think it is important to understand these two views? How does his picture of science differ from views presented in the first chapter of this book?

3. Why did Eiseley choose to divide his essay into three sections? How are the sections related to each other?

4. After all his discussions of science and evolution, why do you think Eiseley chose to end his essay with the anecdote about the sunflower? What point is he trying to make through this story?

5. In his notebooks, Eiseley once wrote, "I am not nearly so interested in what monkey man was derived from as I am in what kind of monkey he is to become." Would you agree or disagree with this statement? Which view of the relationship between science and nature does this statement support?

4 Science and Ethics

Many people consider "ethics" something that belongs in the realm of philosophy courses, to be debated by people who wear tweed jackets and live in ivory towers. But as working scientists know, ethics are a central concern for today's scientific community. Modern technology and discoveries have given scientists immense powers to change the lives of both individuals and communities, and these powers require a strong ethical code of conduct if we are not to witness further abuses of the type seen in Nazi Germany two generations ago.

The readings in this section look at the evolving picture scientists are drawing of ethics. Eliot Marshall explores the complications that ensue when business interests enter the equation, and Gina Maranto looks at one particular area of interest: genetic engineering. Stanley Milgram offers a classic example of the power authority figures can exert over laypeople, and Karl Jaspers concludes with a philosophical examination of science and existentialism. As you read these essays, ask yourself, "What do I mean by ethics? Who has the right to decide what my ethics are? What ethical standards do I have for people in authority? What power do I have to enforce those standards?"

Eliot Marshall

When Commerce and Academe Collide

ELIOT MARSHALL (born 1949) graduated from Harvard in 1971 and took a staff writing job at the *New Republic*. From there he moved to the news staff of *Science*, where his specialties include the American space program, defense technology, and budget issues relating to science and technology. He is the author of *Legalization: A Debate* (1988), which covers the issues involved in the legalization of drugs, and in 1985 held a Knight Fellowship at Stanford. In the last decade he has published nearly three hundred articles, including this one, which appeared in *Science* in 1990.

Marshall notes that conflicts of interest "have moved from specialized journals like *Science* to the front pages of the newspapers. People are getting more interested in science; it's become more of an institution of government than it used to be, and so people are looking at it very closely." He adds, "When I write about subjects like these, which can get very twisty and complex, I try to tell a story, to make people want to read and follow the directions I'm exploring. I want to be entertaining, but also to strive for clarity, so that my readers can get the point without being hit over the head."

Consider a scientist about to embark on a clinical trial of what promises to be a blockbuster drug. He happens to have shares in the company that manufactures the drug. Should he decline to participate? Conduct the tests but declare his potential conflict of interest when he reports the results? Sell the shares? Or assume that his objectivity will overcome any possible bias and carry on regardless?

2 Or take a university researcher who has developed a new biological technique. He sets up a company with funding from a major corporation to exploit his discovery. But the corporation also funds research in the scientist's university lab. How can he respond to a charge that university resources are being used for private gain? And what about the students? How can they be sure that a professor's advice—on such things as choosing a thesis topic—is inspired by academic and not business interests?

3 Questions like these confront and often haunt faculty members and deans these days—a product of the boom in university-industry partnerships. And one reason they haunt academe is that they have also caught the attention of Congress. Last year, two congressional committees put the spotlight on academic conflicts of interest, focusing on cases in which commercial agreements went sour or violated ethical standards.

4 Goaded into action by these hearings, the National Institutes of Health (NIH) drafted a set of rules designed to steer its grantees toward a common approach to the problem. The guidelines would have required faculty to disclose their investments, along with those of their children and spouses, to college administrators. In addition, they would have prohibited faculty from having an interest in any company whose products they were testing. The rules sparked a storm of protest, and they were withdrawn last December (*Science*, 12 January, p. 154).

5 But the federal government hasn't entirely quit the field. There are rumblings that Representative Ted Weiss (D-NY) may attempt to add conflict-of-interest standards to the NIH authorization bill this year. And the Department of Health and Human Services may yet come back with a revised set of rules. Meanwhile, a few schools have taken the moral high ground by voluntarily adopting tighter standards. Their responses are diverse, however, making for a confusing array of requirements that vary from one university to another, even from one department to another. Some codes are quite explicit, like those approved by the Harvard Medical School in March, but others treat the subject almost in hypothetical terms

6 The problem is anything but hypothctical, however. The two situations sketched at the beginning of this article, for example, have real-life counterparts. In 1988 and 1989, it was revealed that investigators at several clinical centers reviewing the heart attack drug

TPA were also stockholders in the company that makes TPA. Several of them later signed a pledge agreeing that in the future they would not hold stock in a company whose product they were studying. And in the early 1980s a controversy broke out among the faculty of the University of California at Davis when it became known that Allied Chemical Company was funding research on nitrogen fixation in plant geneticist Ray Valentine's lab and at Calgene, a company Valentine had founded. Valentine resolved the conflict by dropping out of the university-based research program sponsored by Allied. The university also adopted more stringent rules asking faculty to disclose outside commercial activities and established a standing committee to review potential research conflicts.

7 A similar conflict now confronts George Levy, a chemist at Syracuse University. Levy described his own predicament during a public meeting earlier this year in an attempt to show how the proposed NIH rules would punish innovators. A decade ago, Levy had an idea for improving the software for the nuclear magnetic resonance machines he uses in his research. He advanced the state of the art, and in 1983 founded a small company called New Methods Research, Inc. In 1986, NMRI moved off campus, and by 1988 it had $2.1 million in sales. The following year, NMRI was sold to new owners. Levy returned to his university lab to resume academic research full time. Meanwhile, the university lab got caught in the funding pinch at the National Institutes of Health and discovered this year that it may lose its grant.

8 In desperation, Levy says he may ask NMRI (in which he still has an interest) to spend discretionary funds on research at Syracuse. This would generate new ideas for the company and keep his lab at Syracuse going. The profits would be shared between the school and NMRI.

9 This rescue may succeed, but it makes Levy uncomfortable because, "Here am I, sitting in the middle," trying to negotiate between NMRI and the university. Levy's position is troublesome because he has a direct personal financial stake in the outcome. "I don't like it," he says.

10 Syracuse has been kept informed at every stage of NMRI's creation and development, says Levy, and he is very much in favor of disclosing the necessary details of academic industry deals like this.

11 The university has no formal conflict-of interest guidelines to cover such situations, according to spokesperson Sandi Mulchonry.

"The departments handle it on a case-by-case basis," she says. But Levy says that neither the rescue of his lab nor the creation of NMRI in the first place would have been permissible under NIH's aborted conflict rules. Nor is it likely that the rescue would be allowed under stringent rules being adopted voluntarily today by Harvard and several other universities.

12 Indeed, even a brief survey of major schools reveals, as Carol Scheman of the Association of American Universities says that "there are a huge number of ways in which institutions approach these issues." Some institutions are taking a laissez faire approach. Caltech, for example, relies on its strong honor code to keep the faculty out of conflict situations, according to vice provost David Goodstein. "There are no requirements for disclosure as far as I know," he says. The only "really explicit rule" is that faculty may not take operational responsibilities outside the school. "We have not had any problems," Goodstein says. In contrast, the Massachusetts Institute of Technology requires everyone, staff and faculty alike, to file full outside interest reports every year.

13 The latest to adopt strong measures is Harvard University, and many observers believe its rules could be a model for others. Harvard was stung last year by news coverage of a researcher named Scheffer C. G. Tseng at a Harvard-affiliated eye clinic who had a financial stake in an experimental eye medicine he was testing on patients Before releasing data showing that the medication was ineffective, Tseng cashed in most of his 530,000 shares in a company established to promote the drug. Two other scientists who advised Tseng, one at Harvard and another at Johns Hopkins, also had a financial stake in the company.

14 Medical school dean Daniel C. Tosteson appointed a committee in 1989 to review conflict-of-interest policies and suggest changes. Tosteson did this, he says, not because of any scandal but because Harvard has encouraged its faculty to spread new ideas to the world through commercial agreements, and many gray areas that were vaguely discussed in the old (1983) rules have now become important.

15 Barbara J. McNeil, chair of the department of health care policy, headed the rule-drafting group. Their recommended changes were unveiled before a full faculty meeting in February, where they met a noisy and hostile reception. Opponents, who had bused in scores of doctors from the Massachusetts General Hospital, dominated the

podium. Many felt, as one professor said, that Harvard was "using a cannon to kill a fly." But a month later, McNeil and the reformers won a quiet victory in the select faculty council, which backed the dean with a lop-sided vote of more than 30 to 1. Says council member Leon Eisenberg, "We sensed the world was watching."

16 The new rules require all faculty members to make a full disclosure of their potential conflicts of interest to university administrators at least once a year, and they require researchers to get explicit approval before embarking on studies funded by companies in which they or their families have a financial interest. They also put strictures on faculty involvement in the operations of profit-making companies. The rules will go into effect in May and faculty members will be allowed 6 months to adjust, either by divesting financial holdings or bringing their research into compliance.

17 Tosteson notes that Harvard's approach is "more explicit" than most. Other institutions that are revising their own conflict-of interest standards have been looking carefully at Harvard's new rules, but many schools will find some of the specific provisions too Draconian. That, at least, is the view of the Association of American Universities' Carol Scheman, who says that Harvard, with its network of 14 affiliated clinics and hospitals, has a "unique and difficult problem" in trying to keep tabs on its diverse faculty.

18 Johns Hopkins University went part way toward a stringent code of ethics in November. According to associate dean David Blake, "We didn't get that many disclosures under the old system," so it was revised. "Our assumption is that the problem is mostly one of perceptions, so disclosure is the key." Faculty must report all written agreements involving privately sponsored research and disclose consulting deals that demand more than 26 days a year. Blake himself does not think that clinical research needs special attention because it is already heavily regulated by the federal government. But the medical school does have one "absolute prohibition": you may not own even one share of a corporation that is sponsoring your research at the university.

19 The Stanford University School of Medicine, according to its dean, David Korn, has begun doing some "spot auditing" of the disclosure forms it requires faculty members to submit each year. In addition, Korn says, the review protocol for human subjects has been rewritten to include an extensive series of questions about the finan-

cial involvement of the investigators and their students. This information must be cleared by the human subjects review committee, and the patients must see it, too.

20 Because of the diversity among individual schools, Korn argues that it makes no sense to issue blanket prohibitions for the entire country, as the NIH guidelines attempted to do. Korn himself advocates using the system of Institutional Review Boards that watch over research on human subjects to do a similar job for conflicts of interest. In this approach, each institution would have to assure the government that it had put an effective system into place, subject to spot auditing by some federal supervisor like the NIH. He thinks this would allow for the greatest local autonomy while maintaining high standards. Stanford works under "the philosophy that people are generally decent and behave well," says Korn. "You don't have to tie them up in a bunch of minutiae." Although employees at state universities must work under very strict prohibitions, "rules like those would be anathema on this campus." In general, Korn thinks national policy should avoid detail and give broad, philosophic direction.

21 This is precisely the aim of two major reports issued this spring by university leaders—one drafted by a panel Korn chaired for the Association of Academic Health Centers (AAHC), and the other by a group at the Association of American Medical Colleges (AAMC), headed by Michael Jackson of the George Washington University School of Medicine.

22 Korn's AAHC report, issued on 22 February, traces the boom in academic-industrial collaborations since the 1970s, now encouraged by federal law, and it notes that there are several areas of growing concern. For example, it says the possibilities for conflicts are "legion" in spin-off companies started by faculty members, because the founders live in both the academic and profit-making worlds and control resources and young people's careers in both.

23 An academic's chief loyalty, both reports say, must always be to the university, but they remain a bit vague in the measures they would use to reinforce that loyalty.

24 One point on which all experts seem to agree is the need to disclose potential conflicts in advance. A set of guidelines issued recently by the AAMC says universities ought to develop procedures for full disclosure of financial and professional interests not only for use by the school but to inform "the interested public." In addition, the

AAMC paper says, institutions should review researchers' personal holdings, including those of the immediate family, at least once a year. Questionable cases should be passed up the chain of command to the university president or, better, to a standing committee. Those at odds with the rules "must be handled expeditiously and conclusively," the AAMC panel believes, and "all decisions must be documented."

25 The AAHC goes further, saying that "significant" financial or other relationships, if they raise a potential conflict of interest, should be "fully and accurately disclosed in all speeches, writing, advertising, public communications, or collegial discussions" of sponsored research.

26 These guidelines are new, but others like them have been in existence at major universities for some time—and "honored in the breach," according to C. Kristina Gunsalus, associate vice chancellor for research at the University of Illinois at Champaign. The way to make principles work, Gunsalus says, is to develop a reporting system that will win faculty cooperation and actually do the tedious job of screening and reading the disclosure forms. You must look for trouble, as she does, because "it is extremely difficult for the most honest and upright of scholars to acknowledge their own conflicts for what they are."

27 Representative Weiss says that while he "applauds" the AAMC and AAHC for developing conflict-of-interest guidelines, they do not go far enough. He favors "strong minimum standards for all research institutions." Unless everyone plays by the same rules, Weiss says, "universities that make serious efforts to minimize conflicts of interest could be at a disadvantage in recruiting scientists who enjoy lucrative financial relationships with the private sector."

28 The consensus among those who actively manage faculty conflicts is that one must begin with written forms. They are "the only thing that everybody agrees is absolutely essential," says John Lombardi, the former provost at Johns Hopkins, now president of the University of Florida at Gainesville. "If you actually disclose and write down the relationships you have, the conflict of interest is much easier to discern."

29 Lombardi finds that 95% of the cases turn out to be fine. But "5% are very difficult because they skirt the borders of a conflict of interest. Then you have to do what rulemakers don't like to do: you have to exercise judgment."

30 Both Gunsalus and Lombardi say that when the system works well, it encourages the faculty to venture out into the commercial world, because the responsibility for error—if something goes wrong—falls squarely on the official who gave approval and not on the individual researcher.

Suggestions for Discussion and Writing

1. What is a conflict of interest? Why is it dangerous for researchers to have a conflict of interest? Whose interests are actually jeopardized by such conflicts? Why do you think it is so hard for scientists to separate themselves from conflicts of interest? What are the pressures on them that might lead them into a conflicting situation?

2. Who does Marshall suggest should "police" researchers? What kind of oversight does he think would work best? Based on your reading of Meyer, would you recommend the use of polygraphs to examine researchers' honesty?

3. Oppenheimer argued that scientists belonged in universities, because of the academic freedoms they enjoyed there. What circumstances, according to Marshall, have changed this picture? How do you think Oppenheimer would react to Marshall's arguments?

4. How does Marshall conclude his article? Why do you think he chose to use such a conclusion? What effect does it have on you as a reader?

5. What ethical restraints are put on scientists on your campus? For example, are there policies for the use of human or animal subjects? for preventing fraud? for accepting research funds? How do the scientists on your campus feel about these restraints, if there are any? If not, do they feel there should be? Present your recommendations as a report to the provost of your school.

Gina Maranto

Genetic Engineering: Hype, Hubris, and Haste

GINA MARANTO was born in 1955 and received her M.A. from Johns Hopkins University in creative writing, while holding a Time, Inc. fellowship in science writing. She is a former staff writer for *Discover* who writes frequently about the environment, biology, and national science policy. Now a freelance writer living in Miami, she has published more than 40 articles in the last decade on women's basketball, the poet William Cowper, fishery management, and Michelangelo's frescoes, as well as many reviews in *The New York Times Book Review*. Currently, she is working on a study of environmental poetry. In 1986, she won the National Association of Science Writers' Science in Society award for a *Discover* article on global warming. This article also appeared in *Discover* in 1986.

Maranto told us, "We're really reaching a point where we have to ask where these roads will end; I'm trying to ask questions that will make readers think about some of the consequences of scientific and technological advances. There's no easy way for a writer to do that."

Suddenly, gene technology, a subject that had lately been relegated to the business section of your newspaper, was back on the front page. First, on Feb. 26, it was reported that Advanced Genetic Sciences Inc. (AGS) had illegally tested a genetically altered bacterium by injecting it into fruit and nut trees on the roof of its headquarters in Oakland, Calif. nine months before the Environmental Protection

Agency (EPA) granted the company a permit to test the microbe outside the laboratory. Advanced Genetic's defense was feeble: the microbes hadn't really been released into the environment because they were contained within the trees.

2 Then, on April 3, came more shocking news: without informing anyone, the U.S. Department of Agriculture (USDA) had on Jan. 16 granted the Omaha-based Biologics Corp., a subsidiary of the Tech-America Group Inc., a license to market a genetically engineered vaccine for pseudorabies, a deadly disease that afflicts pigs, cattle, and sheep. In contravention of federal policy, the USDA had also issued a permit for testing the vaccine outside the laboratory. Biologics scientists used the new vaccine on 1,000 three-day-old pigs in four Midwestern states. There's some question whether the company fully explained the implications of the vaccine to the state agriculture departments involved until six months subsequent to the tests. One week after the Biologics story broke, the USDA suspended the license—to get its paperwork in order, it said. On April 22 it was re-issued, whereupon Jeremy Rifkin filed suit. At a congressional hearing on the matter a week later, there was finger pointing all around, after a committee that reviews genetic engineering research at Texas A & M faulted the researchers involved for sidestepping federal rules by performing newly disclosed outdoor tests in 1984.

3 Both cases came to light as a result of investigations by Rifkin, who heads the Foundation on Economic Trends in Washington, D.C. A social and environmental crusader and genetic engineering's most vocal critic, Rifkin received tips from insiders at AGS and a federal agency, secured evidence of their claims, and alerted the press. Subsequent reports that Advanced Genetic had, in the EPA's words, "knowingly falsified" documents, and that the Agriculture Department had essentially formulated its own procedures for handling recombinant DNA (rDNA) were bad news for the advocates of biotechnology.

4 The AGS and USDA episodes occurred as several biotech products were about to move out of the lab and into the marketplace, and they lent weight to charges made over the past few years by a widening circle of critics—including historians and philosophers of science, public interest groups, politicians, and scientists—that the federal government's approach to regulating genetic engineering has been too lighthanded. While agreeing that biotechnology is likely to benefit

society in the long run, these critics question the wisdom of rushing to release live altered microbes before fully considering the environmental consequences.

5 What worries them most is the prospect that biotechnology's products, like those of past technologies, may wreak changes that cannot be foreseen—and perhaps, at first, not even measured. There's no science of predictive ecology, so any assessment of the hazards of gene-spliced organisms is largely speculative. The risk of an Andromeda strain arising is probably minimal, but there's reason for exercising caution. As the eminent biochemist Erwin Chargaff, a professor emeritus at Columbia, said, "You cannot recall a new form of life." Since scores of genetically engineered products are expected to be ready for marketing or field-testing in the next five years, many critics think safeguards must be instituted now, before, as one skeptic says, "a mini Three Mile Island happens."

6 Critics charge, in addition, that those guiding the new industry are ignoring social and political issues. The picture they paint is of an industry beset by financial difficulties and eager to avoid further delays that might worsen its image among investors, and of an administration dedicated to easing the industry's way and giving the public less of a role in making biotechnological decisions. Groups like the Boston-based Committee for Responsible Genetics, whose board of advisers includes more than a dozen leading biologists, have contended for some time that questions concerning the impact of biotechnology on universities and society at large haven't been adequately addressed by either the companies engaged in gene splicing or the government.

7 In the minds of some biotech watchers, the contention that things are moving too rapidly was borne out in the AGS case. In November, against the advice of soil agronomist Martin Alexander of Cornell, One of its most respected advisers, the EPA granted AGS a permit to spray a strawberry patch in Monterey County with a genetically altered strain of *Pseudomonas* bacterium, called ice-minus, that greenhouse tests have shown will forestall frost formation on the plants. (The planned experiment would have been almost identical to a University of California experiment that was stopped by federal judge John Sirica in the spring of 1984 after a suit by Rifkin. Rifkin was backing up a threat to sue the EPA if it allowed any release of an rDNA product before an environmental impact assessment had been

done.) Local residents, alarmed by the prospect of the AGS test, petitioned the Monterey County board of supervisors, which voted twice to block the spraying of the strawberry patch. At a packed public hearing in January, county residents angrily accused AGS, state officials, and the EPA of negligence: no one from the company or the government agencies had informed them of the planned release; instead they learned about it from national television and radio. In February, Rifkin put out the word that AGS had already tested the altered microbe outdoors.

8 After a hasty investigation, the EPA revoked the ice-minus permit and levied the maximum penalty $20,000, for the wrongdoing—which the company admitted to, although it claimed it had not falsified documents, as charged. The agency left open the possibility that AGS could still carry out the test after conducting further safety studies and winning a second permit.

9 Just days after the Advanced Genetic permit was granted, a spokeswoman for the Committee for Responsible Genetics fired off an angry letter to Steven Schatzow, director of the pesticides section of the EPA: "The recent approval sends a signal that the agency has abandoned its initial 'proceed with caution' approach in deference to the wishes of an impatient biotechnology industry."

10 Senator Albert Gore Jr. (D-Tenn.), who's of the opinion that biotechnology shouldn't be "unduly impeded," also registered displeasure. "I think this experiment is almost devoid of significant risk to the environment," he said. "But the procedure used for approving it is totally inadequate, and future proposed experiments will require the sort of information [on how the microbe might spread] that was waived as a requirement this time." Gore later said, "If the administration cannot come up with a workable solution, the Congress may do it for them."

11 Whatever the fate of the Advanced Genetic experiment, the debate over whether genetically engineered organisms should be let out of the lab isn't likely to die soon. Grass roots fear of gene splicing is running high. After the EPA approved ice-minus, picketers carrying NO GENETIC MANIPULATION placards marched in front of AGS headquarters, and opponents in Monterey began drafting a county law covering releases of genetically altered microbes. Missouri environmental groups have voiced concerns about a proposed outdoor test in that state by the Monsanto Corporation of an engineered soil bac-

terium carrying a gene for a protein toxic to insects. And in April Wisconsin dairy farmers, alarmed by predictions that 20 to 30 percent of dairy owners will be forced out of business within three years after the licensing of a hormone that boosts milk production in cows, called on the Food and Drug Administration (FDA) to take a closer look at the drug.

12 Genetic engineering has engendered a certain amount of cultural schizophrenia from the outset. When it was first brought to the public's attention in the early 1970s, the new biology was billed as having the power to solve just about any global problem including pollution, desertification, hunger, and disease. In some respects that view persists. A presidential committee on biotechnology spoke for scientists and businessmen alike when in 1984 it said, "The tremendous potential of biotechnology to contribute to the nation's economy in the near term, and to fulfill society's needs and alleviate its problems in the longer term, makes it imperative that progress in it be encouraged."

13 But if genetic engineering was the stuff of Utopian Studies 101, it was also the stuff of Faust and Frankenstein. James Watson, who shared a Nobel Prize for the discovery of the structure of DNA, has recounted how scientists in the early '70s started to fear what they had found: "We began to ask whether in the process of possibly discovering the power of 'unlimited good,' we might simultaneously be setting the stage for discovering the power of 'unlimited bad.'" After the 1973 Gordon Conferencc on Nucleic Acids, a group of scientists whose work put them at the leading edge of genetics wrote to the National Academy of Sciences urging the formation of a committee to investigate the risks of gene splicing experiments. Published in the journal *Science,* the letter had the desired effect, and in 1974 Paul Berg, the Stanford Nobelist who wound up heading the National Academy's committee, called for an international suspension of certain types of experiments and for the formation of a permanent advisory committee that would institute safeguards. That brought into being the recombinant DNA advisory committee (RAC) of the National Institutes of Health (NIH).

14 In the wake of a widely publicized international meeting on rDNA held in February 1975 at the Asilomar conference center near Monterey, Calif., and of hearings on genetic engineering before the

Senate Subcommittee on Health and Scientific Research that April, editorial pages began to fill with discussions of the new science, focusing on biohazards—especially what might happen if altered *Escherichia coli,* the bacterial species used most often in gene splicing, were to escape from a lab. If a bacterium loaded with a gene for cholera or a cancer-causing virus got into the air or water, would an epidemic sweep the land?

15 Activists, many of whom had learned their tactics in campaigns against the nuclear industry, took to the streets. In the Cambridge (Mass.) City Council, Alfred Vellucci, Cambridge's flamboyant mayor, led choruses of *This Land is Your Land* amid a battle over Harvard's and MIT's gene-splicing labs. Environmental groups throughout the country petitioned for a ban on gene splicing.

16 By the time the public became involved, many scientists had satisfied themselves that their fears were unwarranted. The enfeebled K 12 strains of *E. coli,* they concluded, couldn't survive for long in the human gut, even though it's a natural habitat of wild (i.e., naturally occurring) strains. Researchers became confident that certain simple precautions long followed by microbiologists, like disinfecting experimental cultures before disposing of them, would keep altered bacteria confined.

17 "When the concern emerged about biohazards, scientists were attempting to act responsibly," says Everett Mendelsohn, a Harvard historian of science who has followed genetic engineering from its inception. "They were acting as scientists were expected to act directly after the Vietnam War." The issue was narrow—the debate centered on the immediate effects if K 12 got out of the lab—and almost no one broached larger issues, like the implications of fiddling with nature at such a deep level. The question seemed rapidly resolvable, and, by tackling it, scientists felt they were acting in an ethical manner. When they began to realize that they had exaggerated the danger, says Mendelsohn, "there was a backlash, and they watered the issue way down."

18 Having once gone through the experience of sparking a grass roots movement against their own enterprise, says another historian, who asked not to be named, scientists aren't eager to do it again. "The argument going on today about the ecological safety of field release strikes scientists as being similar to the lab safety argument," he says. "They have a feeling that they've already been through this, and they

won. So they link arms and resist, and don't want to hear the doubts being voiced."

19 There also seems to be a degree of intellectual snobbery among gene splicers. As the élite corps of biology, they tend to think there's little scientific basis to environmental studies, and suspect that ecologists are making a fuss simply because they want to get their hands in the research till.

20 Harsher judgments have been rendered against non-academic critics. By 1977, Watson, the erstwhile voice for caution, had swung around 180° and labeled those still questioning rDNA's safety as "kooks, shits, and incompetents."

21 Such rhetoric seems to some historians an attempt to shift the burden of proof onto those who have doubts about genetic engineering. But as a way of quieting the controversy surrounding gene splicing, it may have backfired. Says one critic, "People haven't bought the notion that this technology is nothing but good, that we don't have to worry about it and should plough ahead."

22 Lack of public confidence in genetic engineering could mean that biotech firms will be tied up in litigation for years. What's needed to avert such a situation, say many observers, is a concerted effort by industry and government to evaluate risks to health and the environment, and to address other public concerns. "If the biotechnology industry is to flourish," Senator Gore says, "it must have the support of the people of this country. The industry won't be able to survive if the public fears it."

23 Gore and his staff have spent a lot of time studying the minutiae of biotechnology to assess its implications and to determine what Congress might do to regulate—but not strangulate—the industry. To that end Gore has convened nearly half a dozen hearings on the subject in the House and Senate since 1977 and fanned interest among his colleagues. The morning of the day the EPA gave approval for the AGS field test he held another session, this time before an ad hoc congressional body called the Environmental and Energy Study Conference.

24 At the hearing, administration officials crossed swords with Rifkin. Henry Miller, special assistant to the commissioner of the FDA, commented on the immediate issue—the impending ice-minus test—by comparing the release of the microbe to dropping yogurt culture into a trash bin in the subway. Orville Bentley, assistant secretary for science and education for the USDA, went further, claiming that

the issue of environmental risk was being overblown; he said that genetically engineered plants and animals didn't differ substantially from those created by traditional hybridization and husbandry. Gore rebuked both witnesses for their "cavalier" attitudes.

25 Predictably, Rifkin again proposed that the government require "total impact statements" from genetic researchers before allowing any releases.

26 Those supporting the industry have tended to lump ecologists with Rifkin—all the better for dismissing the entire lot out of hand. In fact, most ecologists consider Rifkin extreme. Nonetheless, some think he has raised valid points. At the Gore hearing, for instance, he echoed some of the country's leading critics when he charged, "Since World War II our government policy has narrowly focused on the question of potential benefits of new scientific and technological breakthroughs, with little or no attention paid to the long-term environmental, economic, and social costs of new innovations."

27 Most ecologists just want to see the industry properly monitored. In a letter to *Science* last July, five prominent environmental researchers wrote: "We neither doubt the great potential for benefits resulting from the ability to move genes between unrelated species, nor do we believe that most plans for such projects would have severely harmful ecological impacts. We would argue, however, that even traditional breeding has not been ecologically trouble-free, that engineered organisms are analogous to exotic [imported or transplanted] ones to some extent, and that the particular kinds of engineering that are now contemplated are quite likely, if inadequately regulated, to lead to some instances of ecological harm."

28 Molecular biologists have insisted that altered organisms can hardly cause trouble when they're released, because in most cases researchers merely remove, replace, or add one gene in modifying a microbe or plant. Similar mutations continually arise in bacterial populations in the wild without causing problems, they claim.

29 This argument was used by David Kingsbury, the National Science Foundation's assistant director for biological, behavioral, and social sciences, to explain why the animal and plant health inspection service had treated Biologics' pseudorabies vaccine like any other vaccine. He said that since the new vaccine was made simply by deleting one gene from an already approved vaccine, it "didn't fall

within the regulatory guidelines." David Espeseth, an inspection service veterinarian, offered what was in effect the same argument.

30 "The important thing isn't how many genes, but which genes," counters Frances Sharples, an ecologist at Oak Ridge National Laboratory and a member of the NIH's recombinant DNA advisory committee. Indeed, there's ample evidence that single gene changes can significantly alter an organism. For example, certain strains of the bacterium that causes gonorrhea are immune to penicillin because they've acquired a gene that enables them to make an enzyme that breaks down the drug. Among insects, pesticide resistance is also conveyed by one gene.

31 It's clear, ecologists go on to say, that future introductions of genetically engineered organisms may prove as disastrous as past introductions of exotics. Two examples: gypsy moths, imported in 1869 by a Frenchman who wanted to breed disease-resistant worms to produce silk, now cyclically defoliate the Northeast; and the plain brown garden snail from Europe, which was supposed to turn San Jose, Calif. into the escargot capital of the world, has destroyed citrus and vegetable crops in California worth $500 million during a wet and warm winter.

32 What's going to happen when millions of copies of a new microbe are dumped onto a field somewhere? People in the biotech industry are confident that most mutations they create will be feeble compared to wild strains. Of ice-minus, former AGS president Thomas Dyott says, "If the deletion of the ice-nucleating gene gave *Pseudomonas* superior survivability, then mutants lacking the gene should already have taken over."

33 But after a meeting in Philadelphia last summer, a group of microbiologists, ecologists, geneticists, and other scientists, in the first real effort to reconcile their differences, concluded that one can't automatically assume that engineered organisms will be less fit than extant species.

34 At the least, the ability of species to swap genes could render certain biotech projects economically pointless. Several firms, for instance, are working to endow crops with a gene that makes them resistant to broad-spectrum herbicides, which affect not only weeds but also crops that are related to them. If crops were resistant, farmers could douse fields thoroughly without lowering yields. The difficulty is that the crops could breed with the weeds and pass on the trait,

defeating the purpose of the enterprise, not to mention increasing the farmer's woes.

35 At the worst, hybrids might be created that somehow interfered with vital ecological processes such as nitrogen fixation in certain plants, the decay of vegetal matter, or the formation of ice crystals in clouds, which is facilitated by bacteria swept into the atmosphere on dust particles.

36 As yet, there's no way of knowing how likely it is that something of this sort will happen. With limited knowledge of microbial and plant genetics, and minimal understanding of ecosystems, scientists can only guess at the larger import of any release.

37 Nonetheless, some ecologists think the time has come for the first test. The EPA, they say, has done a good job dealing with ice-minus, AGS's indiscretions notwithstanding. It went to the trouble of seeking advice from a panel of non-government scientists, something it's not bound by law to do.

38 Dyott admits that the care EPA officials took in the AGS matter wasn't entirely altruistic. "They knew Rifkin was going to sue," he says, "and they needed to build a strong legal case." There's some indication, in fact, that the agency chose the AGS case because it appeared it would be trouble-free. "We have lots of arguments that indicate that this [ice-minus] will be harmless, and that's really why it's the first case," Dyott said in December. "It's the one the EPA feels most comfortable with."

39 With the exception of Cornell's Alexander, who wanted more tests to determine how likely the microbe was to spread, the EPA's outside advisers concurred that the AGS test should go ahead, because the risks posed by ice-minus were trivial. Ecologists at large seem to have reached a similar conclusion and resigned themselves to the release. "We've come to a point of view that you have to start somewhere," says one ecologist. "The Advanced Genetic test, for God's sake, would cover only two-tenths of an acre."

40 At the National Science Foundation building on Washington's G Street late one rainy afternoon last December, Robert Rabin, assistant director for life sciences in the President's Office of Science and Technology Policy, talked with Kingsbury about why Congress should stay out of the biotechnology business.

41 Kingsbury: You certainly can retard this industry by legislating regulations. The Delaney Clause [which says that drugs showing any evidence of being carcinogenic must be treated as carcinogenic] is an example of how one can run amuck. You're guilty until proved innocent.

42 Rabin: If we apply the same thing to this game, we're going to make it infinitely more difficult not only for industry but also for researchers.

43 Kingsbury: If we approach biotechnology as if it's dangerous until we prove that it's not, we'll never prove it's not, and we'll never go anywhere.*

44 While neither Rabin nor Kingsbury claims that the biotech industry is vital to the U.S. economy, they agree that it's important enough that clearly defined—and none too stringent—regulations should quickly be put in place. Other federal officials have also argued that the government has a duty to help speed genetic engineering along by determining just what's expected in the way of pre-market tests and environmental impact analyses—if not by extending economic incentives. They maintain that the Japanese, the West Germans, the British, and the French have given high priority to biotech, and any bureaucratic tie-ups here could jeopardize the world leadership of U.S. firms. Since biotechnology promises enormous economic and social benefits and poses few risks, they say, there's even more reason not to delay.

45 Many experts dispute such reasoning; they contend that biotechnology's putative benefits are pie in the sky and shouldn't be used to justify an easing of regulations. Some even say it would be no great loss to the world if many of the products now in the pipeline never reached the market.

46 Even considering what an invaluable tool for understanding cells gene splicing is—and there's no denying its usefulness in this regard—a number of critics find it hard to believe that biotechnology can live up to the promises that have been made for it over the years.

---

* In the modern era, regulation may have slowed the march of science and technology, but it hasn't prevented them from going anywhere. For example, the petrochemical industry has been heavily regulated and has nonetheless flourished: it now produces 8,000 chemicals.

47 So far only monoclonal antibodies, tailor-made substances used to treat and diagnose diseases, have deserved their good press: they've been a monetary, but not an unqualified scientific, success. And the rest? "There has been a great deal of hyperbole surrounding biotechnology," says Luther Smithson, director of the biotechnology program at SRI International, a research and consulting firm in Menlo Park, Calif. Much of it was generated when the computer industry was beginning to mature, and biotech fever was centered in the same place—northern California—and at the same institutions as the electronics boom. Smithson says, "Some very bright people got involved in promoting the field, scientists and savvy analysts from investment firms like Hambrecht and Quist who pressed the magic button: cancer." Academics in search of additional sources of funds and financial analysts "looking for excitement made biotechnology their bright star," Smithson adds.

48 How has the bright star shone? As of January nearly 5,700 patent applications for biological processes and products were pending on the desks of examiners in Group 120 of the Patent and Trademark office. Group 120 gets applications having to do with gene splicing, monoclonal antibodies. and bioprocessing, which is essentially the fermentation of microbes. For example, a biological process would be the technique that allows scientists to clone genes. It was granted patent No. 4,468,464 on Aug. 28. 1984, four years after the Supreme Court's decision that new life forms are patentable. Through 1984 the patent office had issued 222 patents related to biotechnology: 188 to American corporations, 21 to individuals, 10 to the U.S. government, and 3 to foreign applicants.

49 Except for the occasional treatment for a plant or animal ailment—Dutch elm disease, say, or channel catfish virus—most of the inventions fall in the human medicine category. The nearly 300 firms dedicated to biotechnology cleave along these same lines. A few companies have staked out crops and cattle, but it will be years before scientists understand the biology of plants and animals well enough to develop many useful products.

50 Most of the companies are left to scramble for a slice of the health-care pie, focusing on a few key substances, especially cancer drugs (like tumor necrosis factor); substances that either boost or impede the functioning of the immune system (alpha, beta, and gamma interferons, and the interleukins); vaccines (for AIDS and hepatitis B);

and hormones (insulin and human growth hormone). By April only two such substances were on sale in the U.S.: human insulin and growth hormone. Two others, alpha and beta interferon, were being sold abroad, where it's easier to get drugs on the market. Estimates vary, but there may be as many as a hundred products climbing through the several stages of clinical trials required by the FDA. According to a report by Paine Webber, another twenty or so were undergoing animal testing in 1985. Some three dozen more were being developed or researched.

51 Halsted Holman, an M.D. who teaches internal medicine at Stanford, offers this analysis of biotechnological products: "Cetus and Genentech [the two biggest biotech companies], they're good at what they do—nobody can deny that. They're coming up with reagents in amounts and purities that were never previously possible. If you happen to be of a reductionist frame of mind, then the availability of these allegedly new products is the beginning of nirvana." Holman, though, puts them in the category of "better tires."

52 "A lot of these first generation products are no great shakes," Smithson says, "not even those for cancer diagnostics or therapeutics." For instance, insulin harvested from pig and cattle pancreases by conventional chemical methods is still cheaper than the genetically engineered version, and there's no shortage of it on the world market.

53 Other products look less impressive than they first did. The interferons and interleukins give physicians only slightly more fire power than conventional cancer therapies. Just one item, says Smithson, looks like a "home run ball"—tissue plasminogen activator (TPA), a substance that dissolves blood clots. A large clinical trial has begun, with results expected around 1990.

54 And a few of the genetically engineered products seem downright useless—bovine growth hormone, for one, which increases milk yields in cows by fifteen to 40 per cent. Says Sheldon Krimsky, an associate professor of urban and environmental policy at Tufts, "There's no crying need for milk in this country. If children go without, it's not because the dairy industry isn't producing enough." Milk shortages exist in Third World countries, but the hormone is too expensive for use there.

55 Has biotechnology been oversold? "At worst, biotechnology has been over*bought*," says Paine Webber's Linda Miller. "The problems genetic engineering is dealing with—cancer and AIDS, to name just

two—are so awful that people wanted to believe biotech would be all things to all conditions. Now investors are seeing it in more realistic terms. There's a higher degree of selectivity and less willingness to go in where a company doesn't have a product in hand or a well focused business plan, or if it has no track record or has suspicious products."

56 Alan Goldhammer, a spokesman for the Industrial Biotechnology Association, says investors made the mistake of seeing biotech through the haze of the computer revolution: "Everybody got really excited and said, 'Let's spend money.' They had seen what happened with microelectronics. But this is a different issue. What the investment community neglected to recognize—and this has now given everybody pause—is that it takes seven years to get a new drug through the FDA."

57 But the delays that have occurred so far can't be pinned on the feds. It has taken nearly a decade for scientists to become adroit at isolating desired genes and to fine-tune the methods for getting the host bacterium or yeast to express genes—that is, to translate the DNA code into protein. Separating the manufactured protein from the bacteria and the culture they're growing in is still more art than science, especially when you have to find the procedure that yields the highest quality at the lowest cost. And the intricacies of scale-up, the switch from fermenting in small containers to fermenting in industrial-sized vats, make the whole enterprise dicy.

58 Although industry insiders like Goldhammer don't see it this way, it appears that progress has been slow mainly because gene splicing moved out of the basic research labs prematurely. Says Smithson, "The first stage of biotechnology has been more important from the basic research standpoint than from the applied."

59 Between 1977 and 1979, eleven bills pertaining to the regulation of biotechnology were introduced in Congress. None passed. Initially, NIH's recombinant DNA committee, which drafted and frequently revised stringent guidelines for experiments, filled the regulatory void. But although universities receiving federal money fell under the committee's aegis, private companies didn't, except by their own choice. While environmentalists focused on safety issues, many social scientists questioned whether the best approach to ensuring that the industry's activities and products were safe was to rely on companies' voluntary compliance with the NIH guidelines.

60 How, or whether, to regulate the industry remains a major issue.

Krimsky divides the opinion into four camps. There's the conservative, free enterprise, don't-regulate-until-you-see-the-whites-of-their-eyes camp—a small minority. "Forget the hazards," they say. "Let's push on." There's the camp that says, "Let's have reasonable regulations and proceed." Then there are the critics that say, "Let's negotiate our demands. Where the process will end, we can't tell. Society will make the ultimate decision." Last, there's the camp that says, "Let's not proceed at all," and will use any legal technique to obstruct the industry's activities. Rifkin is the main and maybe the only true devotee of this school of thought.

61 The Reagan administration falls between the first camp and the second. While it's opposed to the drafting of new laws to regulate biotechnology, of late it has moved toward putting biotech under existing federal statutes, including the Federal Insecticide, Fungicide, and Rodenticide Act, the Toxic Substances Control Act, and the Food, Drug, and Cosmetic Act. A *Federal Register* notice on the last day of 1984 outlined the protocols that the EPA, the FDA, and the USDA intended to follow in handling altered bacteria, proteins, drugs, diagnostics, foods, plant hybrids, and other products created using rDNA techniques, monoclonal antibodies, or bioprocessing. A follow-up notice ran in November 1985. It was some of these protocols, critics say, that were violated by the USDA in the Biologics case.

62 The agencies have dealt with biotech firms based on these provisional notions, though definitions sometimes have to be stretched to fit a particular case. Ice-minus, for example, is being regulated as a pesticide, although it doesn't actually kill the ice-plus *Pseudomonas* strain that promotes frost.

63 In October the Biotechnology Science Coordinating Committee (BSCC) was chartered under the President's Office of Science and Technology Policy to facilitate regulation of the industry. Its stated purpose is to foster consistency and cooperation among the various agencies involved. Before the USDA licensing story broke, Rabin proclaimed, "We're going to see to it that none of the agencies is caught with no clothes on."

64 Part of the role of the BSCC seems to be to allay public fears, which, depending on your view, can be salutary or insidious. Rabin and Kingsbury suggest that the BSCC, of which Kingsbury is chairman, may diminish the power of the NIH's recombinant DNA committee. Although administration officials say the recombinant DNA

panel has no statutory authority and simply couldn't handle the fast-paced biotech industry, others say the real motive for sidestepping it was to move the decision making on genetic engineering further out of the public arena. Says Krimsky, who sat on the recombinant DNA committee from 1979 to 1981, "The BSCC has no public members can proceed with almost no outside input, and offers little public access. It can be seen as nothing more than a system to ease the regulatory process for an industry that's finding it increasingly difficult to gain the confidence of the investment community."

65 Biotech executives now seem resigned to what's happening at the federal level. Two years ago there were those who thought companies could avoid regulation, and those who felt they could live with regulation—if they only knew what form it would take. Now the latter view prevails, and insiders realize they were mistaken to assume that biotechnology would be exempt from supervision. Says Dyott, "When biotechnology was first coming onto the horizon, a lot of people contrasted it with the—in their minds—sinister chemical industry. Since biotechnology was about life processes, presumably its products would be biodegradable. So they thought that biotechnology would be virtually free of regulation. But it was naïve to expect that biotechnology is so clean that no one would have questions about it, and that it wouldn't require regulation."

66 Whether that regulation will be under existing statutes or new ones isn't clear. Despite the failures of comprehensive biotech bills to get through Congress up till now, another move in this direction is being made. Shortly before *Discover* went to press, Representative Don Fuqua (D-Fla.), chairman of the House science and technology committee, introduced a bill that would give the BSCC legal status and make it mandatory for agencies to submit test permits to internal advisory panels. Even without the congressional stamp of approval, Kingsbury was hoping to issue by the end of May blanket policy statements, hammered out by his committee and the pertinent agencies, that he believes will help put the whole debate over regulation to rest. His attitude, reflected in drafts of the policy statements, is that genetically engineered products are no different from those produced by, say, standard chemical means. The slogan he uses is, "We should be looking at products, not processes." Interferon from white blood cells, he says, is the same as interferon from *E. coli.* While this may be true of some blood products, not all genetically engineered products have models

in nature.

67 Members of the camp that wants to negotiate on biotech say that whatever regulatory method is chosen, it must make provisions for public involvement. The Committee for Responsible Genetics advocates public representation at all levels of the review process, which presumably means in the federal agencies as well as on the BSCC—or any similar committee that might be created. Other experts recommend a plural system, with national and local regulatory bodies all having public members. "It would be messy," says Mendelsohn, "but in a democratic society a lot of things are messy. It would give you a way to slow things down so that people would have a chance to ask what values they were trading off—scientific for democratic, democratic for scientific—at each point along the way."

68 In January, in a hotel not far from the Louisiana Superdome, some three hundred scientists were supposed to be engaging in a little futuristic thinking at a conference entitled "*Bio/Technology* Looks to the Next Decade." *Bio/Technology,* the publication that sponsored the get-together, had planned three days of crystal-ball gazing, but the scientists weren't cooperating.

69 The typical talk recounted a company's quest to identify a particular gene, or find a way to purify proteins, or get bacteria or yeast to multiply in 1,500-liter vats. Often elided were critical steps and experimental niceties. When curious colleagues demanded details during question periods, speakers deflected them with rejoinders like "That's proprietary information" or "The lawyers won't let me talk about that." Such comments drew empathetic chuckles from the audience.

70 In the public sessions no new information came out, as it might have at an astrophysics or geology convocation. Nor were there stormy showdowns between high-powered egos, which are the spice of any scientific meeting—what the rank and file yearn for. It's unlikely, too, that the conference-goers exchanged anything of note in private. Said a young researcher from a leading gene-splicing firm, "If someone's worked on a project for two years, he's not going to tell you how he got things to work. Besides, everyone signs a secrecy agreement when he's hired. Maybe if someone got drunk at dinner, he'd give you a clue, but probably not."

71 The lack of candor at the New Orleans conference was an object

lesson: as molecular biologists have moved out of the lab into the board room, the social values of their science have changed. The free exchange of information, some critics argue, has been all but eliminated as researchers have shifted their priorities. A study in progress by Krimsky reveals that at least 362 scientists employed at universities, including 64 who are in the National Academy, sit on the advisory boards of biotech firms. The number holding equity positions or consultancies, or serving on corporate boards, will prove to be several times greater, Krimsky suspects.

72 Such affiliations, says Leon Wofsy, an immunologist in the department of microbiology and immunology at the University of California at Berkeley, go against the old mores of biology. Throughout the fifties and sixties, he says, biologists harbored the notion that their basic research, however arcane it seemed, would ultimately benefit society. They looked down on researchers in the medical field, which was then ascendant because of the federal government's "enormous and unusual commitment to public health," and took a certain pride in being the poor, virtuous cousins. There was even outright hostility toward scientists who attempted to patent research.

73 But around 1975 the potential for commercialization became apparent, and molecular biologists began looking at things differently. The measure of a career had always been the Nobel; suddenly, there was a new criterion of academic success. Scientists like biochemist Walter Gilbert, the Harvard Nobelist who quit academia to head Biogen and made a fortune, at least on paper, became the model. (Gilbert later left Biogen, after it lost $13 million in 1984.)

74 The involvement of scientists with for-profit companies has been further encouraged by fears of dwindling government support for basic research, says Wofsy. Pressure is on researchers to find industrial benefactors, while universities must scramble for a piece of the R & D pie. They dream of deals like the $23.5 million one Harvard struck with Monsanto, giving the company access to faculty expertise and patentable research, or the one Massachusetts General Hospital made with Hoechst of West Germany for $70 million.

75 Many critics question the ethics of such arrangements, because virtually all the advances that made the revolution in biology possible came as a consequence of federal largesse. After footing the bill so long, the public, they say, should share in whatever profits are made.

Moreover, the public should have a say in what direction research takes. "With many aspects of biotechnology having public funding, it's not a question of whether we have public participation," says Mendelsohn, "but of what sort."

76 The thought of intense public scrutiny strikes many scientists as distasteful. Watson once voiced dismay that "public members [of review boards] may take regulation seriously, unlike the molecular biologists." Some critics say this attitude isn't surprising, considering what researchers have at stake. But indications are that if the public were to get involved on a large scale, the progress of research in university or commercial labs wouldn't necessarily be slowed. In 1982 Diana Dutton, a senior research associate at Stanford, surveyed the records of 20 biosafety committees in California that monitored biotechnology experiments at academic and industrial labs. She found that during the year studied, 1979, committees with public members reviewed almost twice as many research proposals as those with none.

77 The position that non-scientists couldn't master enough of the complexities of rDNA and monoclonal antibodies and bioprocessing to do a competent job of assessing the risks—a claim sometimes made by the experts—isn't borne out by events. When people feel threatened, they manage to learn a great deal. "My point of departure is medicine," says Holman. "I have the privilege of seeing real people of all backgrounds addressing themselves to the intrusion of technology into their lives and making decisions about whether they will allow a test, allow a treatment, and so on. There's no question in my mind that average citizens deal with these problems very well, in such a way that I can't argue that they have any less ability to deal with tricky problems than the Paul Bergs."

78 Mendelsohn has a more trenchant reply to experts who say the issues are too tough for the rest of us to comprehend. "If any scientific or technical project is so complicated that the lay public can't understand it when it's explained," he says, "then it shouldn't be done. If it is done, then we've ceded our responsibility for critical decision making to a select élite. No democratic society can tolerate that."

79 Commercialism represents a far more powerful influence on science than even the federal granting system, say historians of science. The invasion of the university by corporations means that the objectivity of scientists is in question—more so than at any time in mem-

ory. Scientists who are tied to the biotech industry, those who have a monetary stake in it, can no longer claim to be disinterested. In evaluating what the technology can do and how much the nation needs it, their word is suspect. "It's the scientists who make the decisions about what the applications are and what risks are acceptable in light of the benefits," says Dutton. "But is that really what society as a whole would decide? I think the evidence indicates that the scientists' record is good in certain technical senses, but not nearly as good as it could be across a whole range of human values that make a society livable, and those are things lay people are expert on—truly expert."

80 At this juncture, however, the issue of who's expert and who's not may be moot. "I don't think the recombinant DNA issue is a big issue," says Holman. "For the most part the battle has been won by the scientific and technical community. University and industry people are united, and forces are too overwhelming to bring back the serious questioning that went on at the beginning of this controversy. From that standpoint, it's all over."

81 He may be right.

Suggestions for Discussion and Writing

1. Why do you think people fear genetic engineering? What are the dangers you hear people mention about this technology? What is an Andromeda strain? Why do opponents of genetic engineering fear one?

2. According to Maranto, what are the potential benefits from bioengineering? Why do researchers think the gene is the place to start altering?

3. What are the social and ethical consequences of using genetically-engineered products? Who currently decides what these consequences are? Who ought to make those decisions? How has the development of genetic engineering changed the "social values of science"?

4. According to Maranto, what role does money play in genetic engineering debates? How and where does she introduce economic factors into her argument? What effect does her money information have on your picture of the overall genetic engineering debate?

5. What is the current status of government approval for genetic engineering? Do you agree with the government's current position? If not, what changes would you suggest?

Stanley Milgram
The Perils of Obedience

STANLEY MILGRAM, born in New York in 1933, was one of America's most controversial behavioral psychologists. After receiving his Ph.D. from Harvard in 1960, he taught at Yale, Harvard, and the City University of New York (CUNY). At CUNY he performed the research into human conformity and aggression for which he was most famous, the results of which were published in 1974 as *Obedience to Authority* (1974). He also wrote *Television and Anti-Social Behavior* with R. Lance Shotland (1973); *Psychology in Today's World* (1975), and *The Individual in a Social World: Essays and Experiments* (1977). Milgram died in 1984.

Obedience to Authority caused a national sensation when it was published; it was used to interpret, among other incidents, the Holocaust, the 1972 My Lai massacres in Vietnam, and the Watergate episode. It was nominated for the National Book Award in 1975. This essay is taken from the book.

Obedience is as basic an element in the structure of social life as one can point to. Some system of authority is a requirement of all communal living, and it is only the person dwelling in isolation who is not forced to respond, with defiance or submission, to the commands of others. For many people, obedience is a deeply ingrained behavior tendency, indeed a potent impulse overriding training in ethics, sympathy, and moral conduct.

2 The dilemma inherent in submission to authority is ancient, as old as the story of Abraham, and the question of whether one should obey when commands conflict with conscience has been argued by

Plato, dramatized in *Antigone,* and treated to philosophic analysis in almost every historical epoch. Conservative philosophers argue that the very fabric of society is threatened by disobedience, while humanists stress the primacy of the individual conscience.

3 The legal and philosophic aspects of obedience are of enormous import, but they say very little about how most people behave in concrete situations. I set up a simple experiment at Yale University to test how much pain an ordinary citizen would inflict on another person simply because he was ordered to by an experimental scientist. Stark authority was pitted against the subjects' strongest moral imperatives against hurting others, and, with the subjects' ears ringing with the screams of the victims, authority won more often than not. The extreme willingness of adults to go to almost any lengths on the command of an authority constitutes the chief finding of the study and the fact most urgently demanding explanation.

4 In the basic experimental design, two people come to a psychology laboratory to take part in a study of memory and learning. One of them is designated as a "teacher" and the other a "learner." The experimenter explains that the study is concerned with the effects of punishment on learning. The learner is conducted into a room, seated in a kind of miniature electric chair; his arms are strapped to prevent excessive movement, and an electrode is attached to his wrist. He is told that he will be read lists of simple word pairs, and that he will then be tested on his ability to remember the second word of a pair when he hears the first one again. Whenever he makes an error, he will receive electric shocks of increasing intensity.

5 The real focus of the experiment is the teacher. After watching the learner being strapped into place, he is seated before an impressive shock generator. The instrument panel consists of thirty lever switches set in a horizontal line. Each switch is clearly labeled with a voltage designation ranging from 15 to 450 volts. The following designations are clearly indicated for groups of four switches, going from left to right: Slight Shock, Moderate Shock, Strong Shock, Very Strong Shock, Intense Shock, Extreme Intensity Shock, Danger: Severe Shock. (Two switches after this last designation are simply marked XXX.)

6 When a switch is depressed, a pilot light corresponding to each switch is illuminated in bright red; an electric buzzing is heard; a blue light, labeled "voltage energizer," flashes; the dial on the voltage

meter swings to the right; and various relay clicks sound off.

7 The upper left-hand corner of the generator is labeled SHOCK GENERATOR, TYPE ZLB, DYSON INSTRUMENT COMPANY, WALTHAM, MASS. OUTPUT 15 VOLTS—450 VOLTS.

8 Each subject is given a sample 45-volt shock from the generator before his run as teacher, and the jolt strengthens his belief in the authenticity of the machine.

9 The teacher is a genuinely naïve subject who has come to the laboratory for the experiment. The learner, or victim, is actually an actor who receives no shock at all. The point of the experiment is to see how far a person will proceed in a concrete and measurable situation in which he is ordered to inflict increasing pain on a protesting victim.

10 Conflict arises when the man receiving the shock begins to show that he is experiencing discomfort. At 75 volts, he grunts; at 120 volts, he complains loudly; at 150, he demands to be released from the experiment. As the voltage increases, his protests become more vehement and emotional. At 285 volts, his response can be described only as an agonized scream. Soon thereafter, he makes no sound at all.

11 For the teacher, the situation quickly becomes one of gripping tension. It is not a game for him; conflict is intense and obvious. The manifest suffering of the learner presses him to quit; but each time he hesitates to administer a shock, the experimenter orders him to continue. To extricate himself from this plight, the subject must make a clear break with authority.

12 The subject, Gretchen Brandt,* is an attractive thirty-one-year-old medical technician who works at the Yale Medical School. She had emigrated from Germany five years before.

13 On several occasions when the learner complains, she turns to the experimenter coolly and inquires, "Shall I continue?" She promptly returns to her task when the experimenter asks her to do so. At the administration of 210 volts, she turns to the experimenter, remarking firmly, "Well, I'm sorry, I don't think we should continue."

14 EXPERIMENTER: The experiment requires that you go on until he has learned all the word pairs correctly.

---

* Names of subjects described in this piece have been changed.

15 BRANDT: He has a heart condition, I'm sorry. He told you that before.

16 EXPERIMENTER: The shocks may be painful but they are not dangerous.

17 BRANDT: Well, I'm sorry, I think when shocks continue like this, they *are* dangerous. You ask him if he wants to get out. It's his free will.

18 EXPERIMENTER: It is absolutely essential that we continue. . . .

19 BRANDT: I'd like you to ask him. We came here of our free will. If he wants to continue I'll go ahead. He told you he had a heart condition. I'm sorry. I don't want to be responsible for anything happening to him. I wouldn't like it for me either.

20 EXPERIMENTER: You have no other choice.

21 BRANDT: I think we are here on our own free will. I don't want to be responsible if anything happens to him. Please understand that.

22 She refuses to go further and the experiment is terminated.

23 The woman is firm and resolute throughout. She indicates in the interview that she was in no way tense or nervous, and this corresponds to her controlled appearance during the experiment. She feels that the last shock she administered to the learner was extremely painful and reiterates that she "did not want to be responsible for any harm to him."

24 The woman's straightforward, courteous behavior in the experiment, lack of tension, and total control of her own action seem to make disobedience a simple and rational deed. Her behavior is the very embodiment of what I envisioned would be true for almost all subjects.

An Unexpected Outcome

25 Before the experiments, I sought predictions about the outcome from various kinds of people—psychiatrists, college sophomores, middle-class adults, graduate students and faculty in the behavioral sciences. With remarkable similarity, they predicted that virtually all subjects would refuse to obey the experimenter. The psychiatrists, specifically, predicted that most subjects would not go beyond 150 volts, when the victim makes his first explicit demand to be freed. They expected that only 4 percent would reach 300 volts, and that only a pathological fringe of about one in a thousand would administer the

highest shock on the board.

26 These predictions were unequivocally wrong. Of the forty subjects in the first experiment, twenty-five obeyed the orders of the experimenter to the end, punishing the victim until they reached the most potent shock available on the generator. After 450 volts were administered three times, the experimenter called a halt to the session. Many obedient subjects then heaved sighs of relief, mopped their brows, rubbed their fingers over their eyes, or nervously fumbled cigarettes. Others displayed only minimal signs of tension from beginning to end.

27 When the very first experiments were carried out, Yale undergraduates were used as subjects, and about 60 percent of them were fully obedient. A colleague of mine immediately dismissed these findings as having no relevance to "ordinary" people, asserting that Yale undergraduates are a highly aggressive, competitive bunch who step on each other's necks on the slightest provocation. He assured me that when "ordinary" people were tested, the results would be quite different. As we moved from the pilot studies to the regular experimental series, people drawn from every stratum of New Haven life came to be employed in the experiment: professionals, white-collar workers, unemployed persons, and industrial workers. *The experiment's total outcome was the same as we had observed among the students.*

28 Moreover, when the experiments were repeated in Princeton, Munich, Rome, South Africa, and Australia, the level of obedience was invariably somewhat *higher* than found in the investigation reported in this article. Thus one scientist in Munich found 85 percent of his subjects obedient.

29 Fred Prozi's reactions, if more dramatic than most, illuminate the conflicts experienced by others in less visible form. About fifty years old and unemployed at the time of the experiment, he has a good-natured, if slightly dissolute, appearance, and he strikes people as a rather ordinary fellow. He begins the session calmly but becomes tense as it proceeds. After delivering the 180-volt shock, he pivots around in his chair and, shaking his head, addresses the experimenter in agitated tones:

30 PROZI: I can't stand it. I'm not going to kill that man in there. You hear him hollering?

31 EXPERIMENTER: As I told you before, the shocks may be painful, but

32 PROZI: But he's hollering. He can't stand it. What's going to happen to him?

33 EXPERIMENTER (his voice is patient, matter-of-fact): The experiment requires that you continue, Teacher.

34 PROZI: Aaah, but, unh, I'm not going to get that man sick in there—know what I mean?

35 EXPERIMENTER: Whether the learner likes it or not, we must go on, through all the word pairs.

36 PROZI: I refuse to take the responsibility. He's in there hollering!

37 EXPERIMENTER: It s absolutely essential that you continue, Prozi.

38 PROZI: (indicating the unused questions): There's too many left here, I mean, Jeez, if he gets them wrong, there's too many of them left. I mean, who's going to take the responsibility if anything happens to that gentleman?

39 EXPERIMENTER: I'm responsible for anything that happens to him. Continue, please.

40 PROZI: All right. (Consults list of words.) The next one's "Slow—walk, truck, dance, music." Answer, please. (A buzzing sound indicates the learner has signaled his answer.) Wrong. A hundred and ninety-five volts. "Dance." (Zzumph!)

41 LEARNER (yelling): Let me out of here! My heart's bothering me! (Teacher looks at experimenter.)

42 EXPERIMENTER: Continue, please.

43 LEARNER (screaming): Let me out of here! You have no right to keep me here! Let me out of here, my heart's bothering me, let me out!

44 PROZI (shakes head, pats the table nervously): You see, he's hollering. Hear that? Gee, I don't know.

45 EXPERIMENTER: The experiment requires. . .

46 PROZI (interrupting): I know it does, sir, but I mean—hunh! He don't know what he's getting in for. He's up to 195 volts! (Experiment continues, through 210 volts, 225 volts, 240 volts, 255 volts, 270 volts, at which point Prozi, with evident relief, runs out of word-pair questions.)

47 EXPERIMENTER: You'll have to go back to the beginning of that page and go through them again until he's learned them all

correctly.

48 PROZI: Aw, no. I'm not going to kill that man. You mean I've got to keep going up with the scale? No sir. He's hollering in there. I'm not going to give him 450 volts.

49 EXPERIMENTER: The experiment requires that you go on.

50 LEARNER: Ohhh. I absolutely refuse to answer anymore. (Shouting urgently now.) Let me out of here! You can't hold me here! Get me out. Get-me-out-of-here!

51 EXPERIMENTER: Continue. The next word is "Green," please.

52 PROZI: "Green—grass, hat, ink, apple." (Nothing happens. No answering buzz. Just gloomy silence.)

53 PROZI: I don't think he is going to answer.

54 EXPERIMENTER: If the learner doesn't answer in a reasonable time, about four or five seconds, consider the answer wrong. And follow the same procedures you have been doing for wrong answers. Say, "Wrong," tell him the number of volts, give him the punishment, read him the correct answer. Continue, please.

55 PROZI: "Green—grass, hat, ink, apple." Answer, please. (More silence. Subject just sits there.)

56 EXPERIMENTER: Go on with the procedure, please.

57 PROZI: Three hundred and fifteen volts. The answer is "ink." (Zzumph!)

58 LEARNER: Ohh. Ooohh.

59 PROZI (relieved to hear response): Next one. . .

60 LEARNER: I absolutely refuse to answer.

61 PROZI: You got to. You get a shock if you don't answer. (Experiment continues with learner refusing to answer, and, finally, failing even to grunt or groan. At 375 volts, the subject stands up.)

62 PROZI: I think something's happened to that fellow in there. I don't get no answer. He was hollering at less voltage. Can't you check in and see if he's all right, please?

63 EXPERIMENTER (with same detached calm): Not once we've started. Please continue, Teacher.

64 PROZI (sits down, sighs deeply): "Cool—day, shade, water, paint." Answer, please. Are you all right in there? Are you all right?

65 EXPERIMENTER: Please continue, Teacher. Continue, please. (Subject pushes lever. Zzumph!)

66 PROZI (swiveling around in his chair): Something's happened to that man in there. (Swiveling back.) Next one. "Low—dollar, neck-

lace, moon, paint." (Turning around again.) Something's happened to that man in there. You better check in on him, sir. He won't answer or nothing.

67 EXPERIMENTER: Continue. Go on, please.

68 PROZI: You accept all responsibility?

69 EXPERIMENTER: The responsibility is mine. Correct. Please go on. (Subject returns to his list, starts running through words as rapidly as he can read them, works through to 450 volts.)

70 PROZI: That's that.

71 EXPERIMENTER: Continue using the 450 switch for each wrong answer. Continue, please.

72 PROZI: But I don't get anything!

73 EXPERIMENTER: Please continue. The next word is "white."

74 PROZI: Don't you think you should look in on him, please?

75 EXPERIMENTER: Not once we've started the experiment.

76 PROZI: What if he's dead in there? (Gestures toward the room with the electric chair.) I mean, he told me he can't stand the shock, sir. I don't mean to be rude, but I think you should look in on him. All you have to do is look in on him. All you have to do is look in the door. I don't get no answer, no noise. Something might have happened to the gentleman in there, sir.

77 EXPERIMENTER: We must continue. Go on, please.

78 PROZI: You mean keep giving him what? Four-hundred-fifty volts, what he's got now?

79 EXPERIMENTER: That's correct. Continue. The next word is "white."

80 PROZI (now at a furious pace): "White—cloud, horse, rock, house." Answer, please. The answer is "horse." Four hundred and fifty volts. (Zzumph!) Next word, "Bag—paint, music, clown, girl." The answer is "paint." Four hundred and fifty volts. (Zzumph!) Next word is "Short—sentence, movie. . ."

81 EXPERIMENTER: Excuse me, Teacher. We'll have to discontinue the experiment.

Peculiar Reactions

82 Morris Braverman, another subject, is a thirty-nine-year-old social worker. He looks older than his years because of his bald head and serious demeanor. His brow is furrowed, as if all the world's burdens were carried on his face. He appears intelligent and concerned.

83 When the learner refuses to answer and the experimenter instructs Braverman to treat the absence of an answer as equivalent to a wrong answer, he takes his instruction to heart. Before administering 300 volts he asserts officiously to the victim, "Mr. Wallace, your silence has to be considered as a wrong answer." Then he administers the shock. He offers halfheartedly to change places with the learner, then asks the experimenter, "Do I have to follow these instructions literally?" He is satisfied with the experimenter's answer that he does. His very refined and authoritative manner of speaking is increasingly broken up by wheezing laughter.

84 The experimenter's notes on Mr. Braverman at the last few shocks are:

> *Almost breaking up now each time gives shock. Rubbing face to hide laughter.*
> *Squinting, trying to hide face with hand, still laughing.*
> *Cannot control his laughter at this point no matter what he does.*
> *Clenching fist, pushing it onto table.*

85 In an interview after the session, Mr. Braverman summarizes the experiment with impressive fluency and intelligence. He feels the experiment may have been designed also to "test the effects on the teacher of being in an essentially sadistic role, as well as the reactions of a student to a learning situation that was authoritative and punitive." When asked how painful the last few shocks administered to the learner were, he indicates that the most extreme category on the scale is not adequate (it read EXTREMELY PAINFUL) and places his mark at the edge of the scale with an arrow carrying it beyond the scale.

86 It is almost impossible to convey the greatly relaxed, sedate quality of his conversation in the interview. In the most relaxed terms, he speaks about his severe inner tension.

87 EXPERIMENTER: At what point were you most tense or nervous?

88 MR. BRAVERMAN: Well, when he first began to cry out in pain, and I realized this was hurting him. This got worse when he just blocked and refused to answer. There was I. I'm a nice person, I think, hurting somebody, and caught up in what seemed a mad situation. . . and in the interest of science, one goes through with it.

89 When the interviewer pursues the general question of tension, Mr. Braverman spontaneously mentions his laughter.

90 "My reactions were awfully peculiar. I don't know if you were watching me, but my reactions were giggly, and trying to stifle laughter. This isn't the way I usually am. This was a sheer reaction to a totally impossible situation. And my reaction was to the situation of having to hurt somebody. And being totally helpless and caught up in a set of circumstances where I just couldn't deviate and I couldn't try to help. This is what got me."

91 Mr. Braverman, like all subjects, was told the actual nature and purpose of the experiment, and a year later he affirmed in a questionnaire that he had learned something of personal importance: "What appalled me was that I could possess this capacity for obedience and compliance to a central idea, i.e., the value of a memory experiment, even after it became clear that continued adherence to this value was at the expense of violation of another value, i.e., don't hurt someone who is helpless and not hurting you. As my wife said, 'You can call yourself Eichmann.' I hope I deal more effectively with any future conflicts of values I encounter."

The Etiquette Of Submission

92 One theoretical interpretation of this behavior holds that all people harbor deeply aggressive instincts continually pressing for expression, and that the experiment provides institutional justification for the release of these impulses. According to this view, if a person is placed in a situation in which he has complete power over another individual, whom he may punish as much as he likes, all that is sadistic and bestial in man comes to the fore. The impulse to shock the victim is seen to flow from the potent aggressive tendencies, which are part of the motivational life of the individual, and the experiment, because it provides social legitimacy, simply opens the door to their expression.

93 It becomes vital, therefore, to compare the subject's performance when he is under orders and when he is allowed to choose the shock level.

94 The procedure was identical to our standard experiment, except that the teacher was told that he was free to select any shock level on any of the trials. (The experimenter took pains to point out that the teacher could use the highest levels on the generator, the lowest, any in between, or any combination of levels.) Each subject proceeded for

thirty critical trials. The learner's protests were coordinated to standard shock levels, his first grunt coming at 75 volts, his first vehement protest at 150 volts.

95 The average shock used during the thirty critical trials was less than 60 volts—lower than the point at which the victim showed the first signs of discomfort. Three of the forty subjects did not go beyond the very lowest level on the board, twenty-eight went no higher than 75 volts, and thirty-eight did not go beyond the first loud protest at 150 volts. Two subjects provided the exception, administering up to 325 and 450 volts, but the overall result was that the great majority of people delivered very low, usually painless, shocks when the choice was explicitly up to them.

96 This condition of the experiment undermines another commonly offered explanation of the subjects' behavior—that those who shocked the victim at the most severe levels came only from the sadistic fringe of society. If one considers that almost two-thirds of the participants fall into the category of "obedient" subjects, and that they represented ordinary people drawn from working, managerial, and professional classes, the argument becomes very shaky. Indeed, it is highly reminiscent of the issue that arose in connection with Hannah Arendt's 1963 book, *Eichmann in Jerusalem.* Arendt contended that the prosecution's effort to depict Eichmann as a sadistic monster was fundamentally wrong, that he came closer to being an uninspired bureaucrat who simply sat at his desk and did his job. For asserting her views, Arendt became the object of considerable scorn, even calumny. Somehow, it was felt that the monstrous deeds carried out by Eichmann required a brutal, twisted personality, evil incarnate. After witnessing hundreds of ordinary persons submit to the authority in our own experiments, I must conclude that Arendt's conception of the banality of evil comes closer to the truth than one might dare imagine. The ordinary person who shocked the victim did so out of a sense of obligation—an impression of his duties as a subject—and not from any peculiarly aggressive tendencies.

97 This is, perhaps, the most fundamental lesson of our study: ordinary people, simply doing their jobs, and without any particular hostility on their part, can become agents in a terrible destructive process. Moreover, even when the destructive effects of their work become patently clear, and they are asked to carry out actions incompatible with fundamental standards of morality, relatively few people have

the resources needed to resist authority.

98 Many of the people were in some sense against what they did to the learner, and many protested even while they obeyed. Some were totally convinced of the wrongness of their actions but could not bring themselves to make an open break with authority. They often derived satisfaction from their thoughts and felt that—within themselves, at least—they had been on the side of the angels. They tried to reduce strain by obeying the experimenter but "only slightly," encouraging the learner, touching the generator switches gingerly. When interviewed, such a subject would stress that he had "asserted my humanity" by administering the briefest shock possible. Handling the conflict in this manner was easier than defiance.

99 The situation is constructed so that there is no way the subject can stop shocking the learner without violating the experimenter's definitions of his own competence. The subject fears that he will appear arrogant, untoward, and rude if he breaks off. Although these inhibiting emotions appear small in scope alongside the violence being done to the learner, they suffuse the mind and feelings of the subject, who is miserable at the prospect of having to repudiate the authority to his face. (When the experiment was altered so that the experimenter gave his instructions by telephone instead of in person, only a third as many people were fully obedient through 450 volts.) It is a curious thing that a measure of compassion on the part of the subject—an unwillingness to "hurt" the experimenter's feelings—is part of those binding forces inhibiting his disobedience. The withdrawal of such deference may be as painful to the subject as to the authority he defies.

Duty Without Conflict

100 The subjects do not derive satisfaction from inflicting pain, but they often like the feeling they get from pleasing the experimenter. They are proud of doing a good job, obeying the experimenter under difficult circumstances. While the subjects administered only mild shocks on their own initiative, one experimental variation showed that, under orders, 30 percent of them were willing to deliver 450 volts even when they had to forcibly push the learner's hand down on the electrode.

101 Bruno Batta is a thirty-seven-year-old welder who took part in the variation requiring the use of force. He was born in New Haven, his parents in Italy. He has a rough-hewn face that conveys a con-

spicuous lack of alertness. He has some difficulty in mastering the experimental procedure and needs to be corrected by the experimenter several times. He shows appreciation for the help and willingness to do what is required. After the 150-volt level, Batta has to force the learner's hand down on the shock plate, since the learner himself refuses to touch it.

102 When the learner first complains, Mr. Batta pays no attention to him. His face remains impassive, as if to dissociate himself from the learner's disruptive behavior. When the experimenter instructs him to force the learner's hand down, he adopts a rigid, mechanical procedure. He tests the generator switch. When it fails to function, he immediately forces the learner's hand onto the shock plate. All the while he maintains the same rigid mask. The learner, seated alongside him, begs him to stop, but with robotic impassivity he continues the procedure.

103 What is extraordinary is his apparent total indifference to the learner; he hardly takes cognizance of him as a human being. Meanwhile, he relates to the experimenter in a submissive and courteous fashion.

104 At the 330-volt level, the learner refuses not only to touch the shock plate but also to provide any answers. Annoyed, Batta turns to him, and chastises him: "You better answer and get it over with. We can't stay here all night." These are the only words he directs to the learner in the course of an hour. Never again does he speak to him. The scene is brutal and depressing, his hard, impassive face showing total indifference as he subdues the screaming learner and gives him shocks. He seems to derive no pleasure from the act itself, only quiet satisfaction at doing his job properly.

105 When he administers 450 volts, he turns to the experimenter and asks, "Where do we go from here, Professor?" His tone is deferential and expresses his willingness to be a cooperative subject, in contrast to the learner's obstinacy.

106 At the end of the session he tells the experimenter how honored he has been to help him, and in a moment of contrition, remarks, "Sir, sorry it couldn't have been a full experiment."

107 He has done his honest best. It is only the deficient behavior of the learner that has denied the experimenter full satisfaction.

108 The essence of obedience is that a person comes to view himself as the instrument for carrying out another person's wishes, and he

therefore no longer regards himself as responsible for his actions. Once this critical shift of viewpoint has occurred, all of the essential features of obedience follow. The most far-reaching consequence is that the person feels responsible *to* the authority directing him but feels no responsibility *for* the content of the actions that the authority prescribes. Morality does not disappear—it acquires a radically different focus: the subordinate person feels shame or pride depending on how adequately he has performed the actions called for by authority.

109 Language provides numerous terms to pinpoint this type of morality: *loyalty, duty, discipline* all are terms heavily saturated with moral meaning and refer to the degree to which a person fulfills his obligations to authority. They refer not to the "goodness" of the person per se but to the adequacy with which a subordinate fulfills his socially defined role. The most frequent defense of the individual who has performed a heinous act under command of authority is that he has simply done his duty. In asserting this defense, the individual is not introducing an alibi concocted for the moment but is reporting honestly on the psychological attitude induced by submission to authority.

110 For a person to feel responsible for his actions, he must sense that the behavior has flowed from "the self." In the situation we have studied, subjects have precisely the opposite view of their actions—namely, they see them as originating in the motives of some other person. Subjects in the experiment frequently said, "If it were up to me, I would not have administered shocks to the learner."

111 Once authority has been isolated as the cause of the subject's behavior, it is legitimate to inquire into the necessary elements of authority and how it must be perceived in order to gain his compliance. We conducted some investigations into the kinds of changes that would cause the experimenter to lose his power and to be disobeyed by the subject. Some of the variations revealed that:

112 • *The experimenter's physical presence has a marked impact on his authority.* As cited earlier, obedience dropped off sharply when orders were given by telephone. The experimenter could often induce a disobedient subject to go on by returning to the laboratory.

113 • *Conflicting authority severely paralyzes action.* When two experimenters of equal status, both seated at the command desk, gave incompatible orders, no shocks were delivered past

the point of their disagreement.

114 • *The rebellious action of others severely undermines authority.* In one variation, three teachers (two actors and a real subject) administered a test and shocks. When the two actors disobeyed the experimenter and refused to go beyond a certain shock level, thirty-six of forty subjects joined their disobedient peers and refused as well.

115 Although the experimenter's authority was fragile in some respects, it is also true that he had almost none of the tools used in ordinary command structures. For example, the experimenter did not threaten the subjects with punishment—such as loss of income, community ostracism, or jail—for failure to obey. Neither could he offer incentives. Indeed, we should expect the experimenter's authority to be much less than that of someone like a general, since the experimenter has no power to enforce his imperatives, and since participation in a psychological experiment scarcely evokes the sense of urgency and dedication found in warfare. Despite these limitations, he still managed to command a dismaying degree of obedience.

116 I will cite one final variation of the experiment that depicts a dilemma that is more common in everyday life. The subject was not ordered to pull the lever that shocked the victim, but merely to perform a subsidiary task (administering the word-pair test) while another person administered the shock. In this situation, thirty-seven of forty adults continued to the highest level on the shock generator. Predictably, they excused their behavior by saying that the responsibility belonged to the man who actually pulled the switch. This may illustrate a dangerously typical arrangement in a complex society: it is easy to ignore responsibility when one is only an intermediate link in a chain of action.

117 The problem of obedience is not wholly psychological. The form and shape of society and the way it is developing have much to do with it. There was a time, perhaps, when people were able to give a fully human response to any situation because they were fully absorbed in it as human beings. But as soon as there was a division of labor things changed. Beyond a certain point, the breaking up of society into people carrying out narrow and very special jobs takes away from the human quality of work and life. A person does not get to see the whole situation but only a small part of it, and is thus unable to act without

some kind of overall direction. He yields to authority but in doing so is alienated from his own actions.

118 Even Eichmann was sickened when he toured the concentration camps, but he had only to sit at a desk and shuffle papers. At the same time the man in the camp who actually dropped Cyclon-b into the gas chambers was able to justify *his* behavior on the ground that he was only following orders from above. Thus there is a fragmentation of the total human act; no one is confronted with the consequences of his decision to carry out the evil act. The person who assumes responsibility has evaporated. Perhaps this is the most common characteristic of socially organized evil in modern society.

Suggestions for Discussion and Writing

1. This is a very long essay, and covers a complex subject. How does Milgram organize the essay to help his readers? What techniques does he use to make it easier for readers to follow his argument?

2. Why is obedience to authority an important ethical concern for scientists? for society? What dilemma does Milgram see as central in the concept of obedience?

3. Why do you think Milgram decided to include the actual transcript of Mr. Prozi's experience, instead of summarizing it? What effect does this transcript have on your reading of the article? Try including actual dialogue in an essay you write. What difference does it make in the effect your writing has on your readers? In what kinds of essays do you think the use of dialogue could have the most impact?

4. What is the importance of terms like *loyalty, duty, discipline* and *obligation* to Milgram's study of authority?

5. Have you ever been in a circumstance where you were ordered to do something you didn't want to do by an authority figure you respected? How did you react? What reasons would you give for reacting in the way you did? Would Milgram's theories have predicted your actions? Assume your answer will be included in an anthology of essays to be published on the twentieth anniversary of the publication of *Obedience to Authority.*

Karl Jaspers

Is Science Evil?

KARL JASPERS was born in Oldenburg, Germany, in 1883. He held various university positions from 1913 to 1948, with the exception of 1937-1945, when he refused to work under the Nazis. From 1948 to his death in 1969, he was associated with the University of Basel, Switzerland. One of the twentieth century's best-known existentialist philosophers, he still considered himself a Christian and believed in a transcendent reality beyond worldly existence. He argued that humans are essentially free, and that each individual determines her or his own fate. "Man is everything," he wrote. "We spend our lives in a seething cauldron of possibilities."

This emphasis on individual freedom was meant to counter what Jaspers saw as the dehumanizing tendencies of modern scientific thought and technological culture. For instance, in *The Future of Mankind* (1957), he argued that humanity was doomed unless nations renounced war as an instrument of policy. His solution was to encourage individuals to reject an "I can't do anything about these problems" attitudes and to encourage them, instead, to take individual action to change government policy. He was the author of over 30 books; this essay originally appeared in *Commentary* in 1950.

No one questions the immense significance of modern science. Through industrial technology it has transformed our existence, and its insights have transformed our consciousness, all this to an extent hitherto unheard of. The human condition throughout the millennia appears relatively stable in comparison with the impetuous movement that has now caught up mankind as a result of science and technology, and is driving it no one knows where. Science has destroyed the substance of many old beliefs and has made others questionable. Its powerful authority has brought more and more men to the point where they wish to know and not believe, where they expect to be helped by science and only by science. The present faith is that scientific understanding can solve all problems and do away with all difficulties.

2 Such excessive expectations result inevitably in equally excessive disillusionment. Science has still given no answer to man's doubts and despair. Instead, it has created weapons able to destroy in a few moments that which science itself helped build up slowly over the years. Accordingly, there are today two conflicting viewpoints: first, the superstition of science, which holds scientific results to be as absolute as religious myths used to be, so that even religious movements are now dressed in the garments of pseudo-science. Second, the hatred of science, which sees it as a diabolical evil of mysterious origin that has befallen mankind.

3 These two attitudes—both non-scientific—are so closely linked that they are usually found together, either in alternation or in an amazing compound.

4 A very recent example of this situation can be found in the attack against science provoked by the trial in Nuremberg of those doctors who, under Nazi orders, performed deadly experiments on human beings. One of the most esteemed medical men among German university professors has accepted the verdict on these crimes as a verdict on science itself, as a stick with which to beat "purely scientific and biological" medicine, and even the modern science of man in general, "this invisible spirit sitting on the prisoner's bench in Nuremberg, this spirit that regards men merely as objects, is not present in Nuremberg alone—it pervades the entire world." And, he adds, if this generalization may be viewed as an extenuation of the crime of the accused doctors, that is only a further indictment of purely scientific medicine.

5 Anyone convinced that true scientific knowledge is possible only of things that *can* be regarded as objects, and that knowledge of the subject is possible only when the subject attains a form of objectivity; anyone who sees science as the one great landmark on the road to truth, and sees the real achievements of modern physicians as derived exclusively from biological and scientific medicine—such a person will see in the above statements an attack on what he feels to be fundamental

to human existence. And he may perhaps have a word to say in rebuttal.

6 In the special case of the crimes against humanity committed by Nazi doctors and now laid at the door of modern science, there is a simple enough argument. Science was not needed at all, but only a certain bent of mind, for the perpetuation of such outrages. Such crimes were already possible millennia ago. In the Buddhist Pali canon, there is the report of an Indian prince who had experiments performed on criminals in order to determine whether they had an immortal soul that survived their corpses: "You shall—it was ordered—put the living man in a tub, close the lid, cover it with a damp hide, lay on a thick layer of clay, put it in the oven and make a fire. This was done. When we knew the man was dead, the tub was drawn forth, uncovered, the lid removed, and we looked carefully inside to see if we could perceive the escaping soul. But we saw no escaping soul." Similarly, criminals were slowly skinned alive to see if their souls could be observed leaving their bodies. Thus there were experiments on human beings before modern science.

7 Better than such a defense, however, would be a consideration of what modern science really genuinely is, and what its limits are.

8 Science, both ancient and modern, has, in the first place, three indispensable characteristics:

9 First, it is *methodical* knowledge. I know something scientifically only when I also know the method by which I have this knowledge, and am thus able to ground it and mark its limits.

10 Second, it is *compellingly certain.* Even the uncertain—i.e., the probable or improbable—I know scientifically only insofar as I know it clearly and compellingly as such, and know the degree of its uncertainty.

11 Third, it is *universally valid.* I know scientifically only what is identically valid for every inquirer. Thus scientific knowledge spreads over the world and remains the same. Unanimity is a sign of universal validity. When unanimity is not attained, when there is a conflict of schools, sects, and trends of fashion, then universal validity becomes prophetic.

12 This notion of science as methodical knowledge, compellingly certain, and universally valid, was long ago possessed by the Greeks. Modern science has not only purified this notion; it has also transformed it: a transformation that can be described by saying that mod-

ern science is *indifferent to nothing.* Everything—the smallest and meanest, the furthest and strangest—that is in any way and at any time *actual,* is relevant to modern science, simply because it *is.* Modern science wants to be thoroughly universal, allowing nothing to escape it. Nothing shall be hidden, nothing shall be silent, nothing shall be a secret.

13 In contrast to the science of classical antiquity, modern science is *basically unfinished.* Whereas ancient science had the appearance of something completed, to which the notion of progress was not essential, modern science progresses into the infinite. Modern science has realized that a finished and total world-view is scientifically impossible. Only when scientific criticism is crippled by making particulars absolute can a closed view of the world pretend to scientific validity—and then it is a false validity. Those great new unified systems of knowledge—such as modern physics—that have grown up in the scientific era, deal only with single aspects of reality. And reality as a whole has been fragmented as never before; whence the openness of the modern world in contrast to the closed Greek cosmos.

14 However, while a total and finished world-view is no longer possible to modern science, the idea of a unity of the sciences has now come to replace it. Instead of the cosmos of the world, we have the cosmos of the sciences. Out of dissatisfaction wilth all the separate bits of knowledge is born the desire to unite all knowledge. The ancient sciences remained dispersed and without mutual relations. There was lacking to them the notion of a concrete totality of science. The modern sciences, however, seek to relate themselves to each other in every possible way.

15 At the same time the modern sciences have increased their claims. They put a low value on the possibilities of speculative thinking, they hold thought to be valid only as part of definite and concrete knowledge, only when it has stood the test of verification and thereby become infinitely modified. Only superficially do the modern and the ancient atomic theories seem to fit into the same theoretical mold. Ancient atomic theory was applied as a plausible interpretation of common experience; it was a statement complete in itself of what might possibly be the case. Modern atomic theory has developed through experiment, verification, refutation: that is, through an incessant transformation of itself in which theory is used not as an end in

itself but as a tool of inquiry. Modern science, in its questioning, pushes to extremes. For example: the rational critique of appearance (as against reality) was begun in antiquity, as in the concept of perspective and its application to astronomy, but it still had some connection with immediate human experiences; today, however, this same critique, as in modern physics for instance, ventures to the very extremes of paradox, attaining a knowledge of the real that shatters any and every view of the world as a closed and complete whole.

16 So it is that in our day a scientific attitude has become possible that addresses itself inquisitively to everything it comes across, that is able to know what it knows in a clear and positive way, that can distinguish between the known and the unknown, and that has acquired an incredible mass of knowledge. How helpless was the Greek doctor or the Greek engineer! The ethos of modern science is the desire for reliable knowledge based on dispassionate investigation and criticism. When we enter its domain we feel as though we were breathing pure air, and seeing the dissolution of all vague talk, plausible opinions, haughty omniscience, blind faith.

17 But the greatness and the limitations of science are inseparable. It is characteristic of the greatness of modern science that it comprehends its own limits:

18 (1) Scientific, objective knowledge is not knowledge of Being. This means that scientific knowledge is particular, not general, that it is directed toward specific objects, and not toward Being itself. Through knowledge itself, science arrives at the most positive recognition of what it does *not* know.

19 (2) Scientific knowledge or understanding cannot supply us with the aims of life. It cannot lead us. By virtue of its very clarity it directs us elsewhere for the sources of our life, our decisions, our love.

20 (3) Human freedom is not an object of science, but is the field of philosophy. Within the purview of science there is no such thing as liberty.

21 These are clear limits, and the person who is scientifically minded will not expect from science what it cannot give. Yet science has become, nevertheless, the indispensable element of all striving for truth, it has become the premise of philosophy and the basis in general for whatever clarity and candor are today possible. To the extent that it succeeds in penetrating all obscurities and unveiling all secrets, science directs us to the most profound, the most genuine secret.

22 The unique phenomenon of modern science, so fundamentally different from anything in the past, including the science of the Greeks, owes its character to the many sources that were its origin; and these had to meet together in Western history in order to produce it.

23 One of these sources was Biblical religion. The rise of modern science is scarcely conceivable without its impetus. Three of the motives that have spurred research and inquiry seem to have come from it.

24 (1) The ethos of Biblical religion demanded truthfulness at all costs. As a result, truthfulness became a supreme value, and at the same time was pushed to the point where it became a serious problem. The truthfulness demanded by God forbade making the search for knowledge a game or amusement, an aristocratic leisure activity. It was a serious affair, a calling in which everything was at stake.

25 (2) The world is the creation of God. The Greeks knew the cosmos as that which was complete and ordered, rational and regular, eternally subsisting. All else was nothing, merely material, not knowable and not worth knowing. But if the world is the creation of God, then everything that exists is worth knowing, just because it is God's creation; there is nothing that ought not to be known and comprehended. To know is to reflect upon God's thought. And God as creator is—in Luther's words—present even in the bowels of a louse.

26 The Greeks remained imprisoned in their closed world-view, in the beauty of their rational cosmos, in the logical transparency of the rational whole. Not only Aristotle and Democritus, but Thomas Aquinas and Descartes, too, obey this Greek urge, so paralyzing to the spirit of science, toward a closed universe. Entirely different is the new impulse to unveil the totality of creation. Out of this there arises the pursuit through knowledge of that reality which is not in accord with previously established laws. In the Logos itself [the Word, Reason] there is born the drive toward repeated self-destruction—not as self-immolation, but in order to arise again and ever again in a process that is to be continued infinitely. This science springs from a Logos that does not remain closed within itself, but is open to an anti-Logos which it permeates by the very act of subordinating itself to it. The continuous, unceasing reciprocal action of theory and experiment is the simple and great example and symbol of the universal process that is the dialectic between Logos and anti-Logos.

27 This new urge for knowledge sees the world no longer as simply

beautiful. This knowledge ignores the beautiful and the ugly, the good and the wicked. It is true that in the end, *omne ens est bonum* [all Being is good], that is, as a creation of God. This goodness, however, is no longer the transparent and self-sufficient beauty of the Greeks. It is present only in the love of all existent things as created by God, and it is present therefore in our confidence in the significance of inquiry. The knowledge of the createdness of all worldly things replaces indifference in the face of the flux of reality with limitless questioning and insatiable spirit of inquiry.

28 But the world that is known and knowable is, as created Being, Being of the second rank. For the world is unfathomable, it has its ground in another, a Creator, it is not self-contained and it is not containable by knowledge. The Being of the world cannot be comprehended as definitive, absolute reality, but points always to another.

29 The idea of creation makes worthy of love whatever is, for it is God's creation; and it makes possible, by this, an intimacy with reality never before attained. But at the same time it gives evidence of the incalculable distance from that Being which is not merely created Being but Being itself, God.

30 (3) The reality of this world is full of cruelty and horror for men. "That's the way things are," is what man must truthfully say. If, however, God is the world's creator, then he is responsible for his creation. The question of justifying God's ways becomes with Job a struggle with the divine for the knowledge of reality. It is a struggle against God, for God. God's existence is undisputed and just because of this the struggle arises. It would cease if faith were extinguished.

31 This God, with his unconditional demand for truthfulness, refuses to be grasped through illusions. In the Bible, he condemns the theologians who wish to console and comfort Job with dogmas and sophisms. This God insists upon science, whose content always seems to bring forth an indictment of him. Thus we have the adventure of knowledge, the furtherance of unrestricted knowledge—and at the same time, a timidity, an awe in the face of it. There was an inner tension to be observed in many scientists of the past century, as if they heard: God's will is unconfined inquiry, inquiry is in the service of God—and at the same time: it is an encroachment on God's domain, all shall not be revealed.

32 This struggle goes hand in hand with the struggle of the man of

science against all that he holds most dear, his ideals, his beliefs; they must be proven, newly verified, or else transformed. Since God could not be believed in if he were not able to withstand all the questions arising from the facts of reality, and since the seeking of God involves the painful sacrifice of all illusions, so true inquiry is the struggle against all personal desires and expectations.

33 This struggle finds its final test in the struggle of the scientist with his own theses. It is the determining characteristic of the modern scientist that he seeks out the strongest points in the criticism of his opponents and exposes himself to them. What in appearance is self-destructiveness becomes, in this case, productive. And it is evidence of a degradation of science when discussion is shunned or condemned, when men imprison themselves and their ideas in a milieu of like-minded savants and become fanatically aggressive to all outside it.

34 That modern science, like all things, contains its own share of corruption, that men of science only too often fail to live up to its standards, that science can be used for violent and criminal ends, that man will steal, plunder, abuse, and kill to gain knowledge—all this is no argument against science.

35 To be sure, science as such sets up no barriers. As science, it is neither human nor inhuman. So far as the well-being of humanity is concerned, science needs guidance from other sources. Science in itself is not enough—or should not be. Even medicine is only a scientific means, serving an eternal ideal, the aid of the sick and the protection of the healthy.

36 When the spirit of a faithless age can become the cause of atrocities all over the world, then it can also influence the conduct of the scientist and the behavior of the physician, especially in those areas of activity where science itself is confused and unguided. It is not the spirit of science but the spirit of its vessels that is depraved. Count Keyserling's dictum—"The roots of truth-seeking lie in primitive aggression"—is as little valid for science as it is for any genuine truth-seeking. The spirit of science is in no way primarily aggressive, but becomes so only when truth is prohibited; for men rebel against the glossing over of truth or its suppression.

37 In our present situation the task is to attain to that true science which knows what it knows at the same time that it knows what it cannot know. This science shows us the ways to the truth that are the indis-

pensable precondition of every other truth. We know what Mephistopheles knew when he thought he had outwitted Faust:

> *Verachte nur Vernunft und Wissenschaft*
> *Des Menschen allerhöchste Kraft*
> *So habe ich Dich schon unbedingt.*
> (Do but scorn Reason and Science
> Man's supreme strength
> Then I'll have you for sure.)

Suggestions for Discussion and Writing

1. Aristotle wrote that "The ultimate end. . . is not knowledge, but action. To be half right on time may be more important than to obtain the whole truth too late." How would Jaspers react to this statement? How do you react to it?

2. What forces in our culture lead non-scientists to believe that "scientific understanding can solve all problems and do away with all difficulties"? How might such attitudes lead to ethical conflicts? Do Marshall or Maranto give you any help in answering this question?

3. How does Jaspers reconcile religious faith with scientific belief? Can you accept his explanation? Would Eiseley? Why or why not?

4. Jaspers uses several numbered lists in his argument. How do these "one-two-three" arguments affect you as a reader? What kind of impression of Jasper's knowledge and authority do they convey to you?

5. How would you answer the question Jaspers poses in his title? What reasons and examples would you use to support your answer?

5 The Costs and Benefits of Medical Technology

If money grew on trees, and everyone had plenty of trees, perhaps we wouldn't worry about the costs and benefits of medical technology. But money is a limited resource, and, as the essays in this section remind us, "costs" can be more than financial. When we weigh the effects of new advances in medical technology, then, we must consider not only if these advances are justified financially but if they fulfill both our individual and community standards for improving life.

Just how complicated this balancing act is can be seen in the readings in this section. Ethicist Sissela Bok begins with a comprehensive examination of the use of placebos in medical care and research. Charles Mee looks at the controversies involved in widespread vaccination. LeRoy Walters looks at the special issues involved with studying and treating AIDS, and Ann E. Weiss reviews the rights patients can expect. Finally, Lester C. Thurow asks a hard question: how much are we willing to spend to prolong life? As you read these essays, ask yourself, "How would I decide the 'worth' of a medical innovation or treatment? What limits would I set? How might I ration health care resources? Can I look beyond my own interests to see the needs of larger communities as I make these decisions?"

Sissela Bok

Placebos

Born in 1934 in Stockholm, SISSELA BOK studied at the Sorbonne in Paris, and at Harvard, where she taught from 1972 until 1984. She now teaches at Brandeis. She serves on numerous committees and advisory boards dealing with medical ethics. Her books include *The Dilemmas of Euthanasia* (with John A. Behnke, 1975); *Lying: Moral Choice in Public and Private Life* (1978); *Secrets: On the Ethics of Concealment and Revelation* (1983); and *A Strategy for Peace: Human Values* (1989).

Bok has long been a campaigner against deception. Recently, she cautioned against American government's increasing reliance on withholding information in an interview with *U.S. News*: "We have been for the world a beacon of open government, and we have to be careful not to squander that leadership." This essay is a chapter from *Lying*.

The common practice of prescribing placebos to unwitting patients illustrates the two miscalculations so common to minor forms of deceit: ignoring possible harm and failing to see how gestures assumed to be trivial build up into collectively undesirable practices. Placebos have been used since the beginning of medicine. They can be sugar pills, salt-water injections—in fact, any medical procedure which has no specific effect on a patient's condition, but which can have powerful psychological effects leading to relief from symptoms such as pain or depression.

2 Placebos are prescribed with great frequency. Exactly how often cannot be known, the less so as physicians do not ordinarily talk publicly about using them. At times, self-deception enters in on the part of physicians, so that they have unwarranted faith in the powers of what

can work only as a placebo. As with salesmanship, medication often involves unjustified belief in the excellence of what is suggested to others. In the past, most remedies were of a kind that, unknown to the medical profession and their patients, could have only placebic benefit, if any.

3 The derivation of "placebo," from the Latin for "I shall please," gives the word a benevolent ring, somehow placing placebos beyond moral criticism and conjuring up images of hypochondriacs whose vague ailments are dispelled through adroit prescriptions of beneficent sugar pills. Physicians often give a humorous tinge to instructions for prescribing these substances, which helps to remove them from serious ethical concern. One authority wrote in a pharmacological journal that the placebo should be given a name previously unknown to the patient and preferably Latin and polysyllabic, and added:

> [I]t is wise if it be prescribed with some assurance and emphasis for psychotherapeutic effect. The older physicians each had his favorite placebic prescriptions—one chose tincture of Condurango, another the Fluidextract of *Cimicifuga nigra.*

4 After all, health professionals argue, are not placebos far less dangerous than some genuine drugs? And more likely to produce a cure than if nothing at all is prescribed? Such a view was expressed in a letter to *The Lancet*:

> Whenever pain can be relieved with a ml of saline, why should we inject an opiate? Do anxieties or discomforts that are allayed with starch capsules require administration of barbiturate, diazepam, or propoxyphene?

5 Such a simplistic view conceals the real costs of placebos, both to individuals and to the practice of medicine. First, the resort to placebos may actually prevent the treatment of an underlying, undiagnosed problem. And even if the placebo "works," the effect is often shortlived; the symptoms may recur, or crop up in other forms. Very often, the symptoms of which the patient complains are bound to go away by themselves, sometimes even from the mere contact with a health professional. In those cases, the placebo itself is unnecessary; having recourse to it merely reinforces a tendency to depend on pills or treatments where none is needed.

6 In the aggregate, the costs of placebos are immense. Many millions of dollars are expended on drugs, diagnostic tests, and psychotherapies of a placebic nature. Even operations can be of this nature—a hysterectomy may thus be performed, not because the condition of the patient requires such surgery, but because she goes from one doctor to another seeking to have the surgery performed, or because she is judged to have a great fear of cancer which might be alleviated by the very fact of the operation.

7 Even apart from financial and emotional costs and the squandering of resources, the practice of giving placebos is wasteful of a very precious good: the trust on which so much in the medical relationship depends. The trust of those patients who find out they have been duped is lost, sometimes irretrievably. They may then lose confidence in physicians and even in bona fide medication which they may need in the future. They may obtain for themselves more harmful drugs or attach their hopes to debilitating fad cures.

8 The following description of a case where a placebo was prescribed reflects a common approach:

> A seventeen-year-old girl visited her pediatrician, who had been taking care of her since infancy. She went to his office without her parents, although her mother had made the appointment for her over the telephone. She told the pediatrician that she was very healthy, but that she thought she had some emotional problems. She stated that she was having trouble sleeping at night, that she was very nervous most of the day. She was a senior in high school and claimed she was doing quite poorly in most of her subjects. She was worried about what she was going to do next year. She was somewhat overweight. This, she felt, was part of her problem. She claimed she was not very attractive to the opposite sex and could not seem to "get boys interested in me " She had a few close friends of the same sex.
>
> Her life at home was quite chaotic and stressful. There were frequent battles with her younger brother, who was fourteen, and with her parents. She claimed her parents were always "on my back." She described her mother as extremely rigid and her father as a disciplinarian who was quite old-fashioned in his values.
>
> In all, she spent about twenty minutes talking with her pediatrician. She told him that what she thought she really needed was tranquilizers, and that that was the reason she came. She felt that this was an extremely difficult year for her, and if she could have something to

calm her nerves until she got over her current crises, everything would go better.

The pediatrician told her that he did not really believe in giving tranquilizers to a girl of her age. He said he thought it would be a bad precedent for her to establish. She was very insistent, however, and claimed that if he did not give her tranquilizers, she would "get them somehow." Finally, he agreed to call her pharmacy and order medication for her nerves. She accepted graciously. He suggested that she call him in a few days to let him know how things were going. He also called her parents to say that he had a talk with her and was giving her some medicine that might help her nerves.

Five days later, the girl called the pediatrician back to say that the pills were really working well. She claimed that she had calmed down a great deal, that she was working things out better with her parents, and had a new outlook on life. He suggested that she keep taking them twice a day for the rest of the school year. She agreed.

A month later, the girl ran out of pills and called her pediatrician for a refill. She found that he was away on vacation. She was quite distraught at not having any medication left, so she called her uncle who was a surgeon in the next town. He called the pharmacy to renew her pills and, in speaking to the druggist, found out that they were only vitamins. He told the girl that the pills were only vitamins and that she could them over the counter and didn't really need him to refill them. The girl became very distraught, feeling that she had been deceived and betrayed by her pediatrician. Her parents, when they heard, commented that they thought the pediatrician was "very clever."

9 The patients who do *not* discover the deception and are left believing that a placebic remedy has worked may continue to rely on it under the wrong circumstances. This is especially true with drugs such as antibiotics, which are sometimes used as placebos and sometimes for their specific action. Many parents, for example, come to believe that they must ask for the prescription of antibiotics every time their child has a fever or a cold. The fact that so many doctors accede to such requests perpetuates the dependence of these families on medical care they do not need and weakens their ability to cope with health problems. Worst of all, those children who cannot tolerate antibiotics may have severe reactions, sometimes fatal, to such unnecessary medication.

10 Such deceptive practices, by their very nature, tend to escape the normal restraints of accountability and can therefore spread more easily than others. There are many instances in which an innocuous seem-

ing practice has grown to become a large-scale and more dangerous one. Although warnings against the "entering wedge" are often rhetorical devices, they can at times express justifiable caution; especially when there are great pressures to move along the undesirable path and when the safeguards are insufficient.

11 In this perspective, there is much reason for concern about placebos. The safeguards against this practice are few or nonexistent—both because it is secretive in nature and because it is condoned but rarely carefully discussed in the medical literature. And the pressures are very great, and growing stronger, from drug companies, patients eager for cures, and busy physicians, for more medication, whether it is needed or not. Given this lack of safeguards and these strong pressures, the use of placebos can spread in a number of ways.

12 The clearest danger lies in the gradual shift from pharmacologically inert placebos to more active ones. It is not always easy to distinguish completely inert substances from somewhat active ones and these in turn from more active ones. It may be hard to distinguish between a quantity of an active substance so low that it has little or no effect and quantities that have some effect. It is not always clear to doctors whether patients require an inert placebo or possibly a more active one, and there can be the temptation to resort to an active one just in case it might also have a specific effect. It is also much easier to deceive a patient with a medication that is known to be "real" and to have power. One recent textbook in medicine goes so far as to advocate the use of small doses of effective compounds as placebos rather than inert substances—because it is important for both the doctor and the patient to believe in the treatment! This shift is made easier because the dangers and side effects of active agents are not always known or considered important by the physician.

13 Meanwhile, the number of patients receiving placebos increases as more and more people seek and receive medical care and as their desire for instant, push-button alleviation of symptoms is stimulated by drug advertising and by rising expectations of what science can do. The use of placebos for children grows as well, and the temptations to manipulate the truth are less easily resisted once such great inroads have already been made.

14 Deception by placebo can also spread from therapy and diagnosis to experimentation. Much experimentation with placebos is honest and consented to by the experimental subjects, especially since the ad-

vent of strict rules governing such experimentation. But grievous abuses have taken place where placebos were given to unsuspecting subjects who believed they had received another substance. In 1971, for example, a number of Mexican-American women applied to a family-planning clinic for contraceptives. Some of them were given oral contraceptives and others were given placebos, or dummy pills that looked like the real thing. Without fully informed consent, the women were being used in an experiment to explore the side effects of various contraceptive pills. Some of those who were given placebos experienced a predictable side effect—they became pregnant. The investigators neither assumed financial responsibility for the babies nor indicated any concern about having bypassed the "informed consent" that is required in ethical experiments with human beings. One contented himself with the observation that if only the law had permitted it, he could have aborted the pregnant women!

15 The failure to think about the ethical problems in such a case stems at least in part from the innocent-seeming white lies so often told in giving placebos. The spread from therapy to experimentation and from harmlessness to its opposite often goes unnoticed in part *because* of the triviality believed to be connected with placebos as white lies. This lack of foresight and concern is most frequent when the subjects in the experiment are least likely to object or defend themselves; as with the poor, the institutionalized, and the very young.

16 In view of all these ways in which placebo usage can spread, it is not enough to look at each incident of manipulation in isolation, no matter how benevolent it may be. When the costs and benefits are weighed, not only the individual consequences must be considered, but also the cumulative ones. Reports of deceptive practices inevitably leak out, and the resulting suspicion is heightened by the anxiety which threats to health always create. And so even the health professionals who do not mislead their patients are injured by those who do; the entire institution of medicine is threatened by practices lacking in candor, however harmless the results may appear in some individual cases.

17 This is not to say that all placebos must be ruled out; merely that they cannot be excused as innocuous. They should be prescribed but rarely, and only after a careful diagnosis and consideration of non-deceptive alternatives; they should be used in experimentation only after subjects have consented to their use.

Suggestions for Discussion and Writing

1. What is a placebo? When do scientists use it? What advantages does a research study using placebos have? How would Bok react to such studies?

2. Reread the anecdote about the seventeen year-old girl who was prescribed a placebo. Two doctors behaved very differently with regard to that prescription. Which doctor do you believe behaved correctly? Why? What would you have done if you had been in each doctor's position?

3. How do you think Bok would feel about the use of lie detectors? On what points might she and Meyer agree? Where might they disagree?

4. Bok has campaigned against lying in any form for many years. Do you agree that the use of placebos is always lying? Is there such a thing as a "good" lie? How would Bok respond to your opinion?

5. What is the role of informed consent in the use of placebos? Why might researchers be tempted to forego informing patients about receiving placebos? Can such omissions ever be ethically justified? Assume you have been appointed to the panel on your campus that oversees research using human subjects. How would you respond if a researcher proposed a study using placebos?

Charles L. Mee, Jr.

The Summer Before Salk

Born in Evanston, Illinois, CHARLES L. MEE, JR. graduated from Howard University in 1960 and shortly thereafter landed a position with *Horizon* magazine, where he served as editor. He has long been associated as well with Off-Broadway theater; he is a published playwright, and has produced a number of Off-

Broadway shows. Among his books are *White Robe, Black Robe* (1973), *Rembrandt's Portrait: A Biography* (1988), and a number of studies of episodes of American history for both adults and juveniles.

In 1983, to celebrate its fiftieth anniversary, *Esquire* commissioned essays about fifty Americans who had changed the world. Mee offered the following contribution to that volume.

The first symptom was the ache and stiffness in the lower back and neck. Then general fatigue. A vaguely upset stomach. A sense of dissociation. Fog closing in. A ringing in the ears. Dull, persistent aching in the legs. By then the doctor would have been called, the car backed out of the garage for the trip to the hospital; by then the symptoms would be vivid: fierce pain, as though the nerves in every part of the body were being probed by a dentist's device without Novocain. All this took a day, twenty-four hours.

2 At the hospital, nurses would command the wheelchair—crowds in the hallway backing against the walls as the group panic made its way down the hall to the examining room, where, amid a turmoil of interns, orderlies, and nurses, the head nurse would step up and pronounce instantly, with authority, "This boy has polio," and the others would draw back, no longer eager to examine the boy, as he was laid out on a cart and wheeled off to the isolation ward while all who had touched him washed their hands.

3 Poliomyelitis is a disease caused by a viral agent that invades the body by way of the gastrointestinal tract, where it multiplies and, on rare occasions, travels via blood and/or nervous pathways to the central nervous system, where it attacks the motor neurons of the spinal cord and part of the brain. Motor neurons are destroyed. Muscle groups are weakened or destroyed. A healthy fifteen-year-old boy of 160 pounds might lose seventy or eighty pounds in a week.

4 As long ago as the turn of the century doctors agreed that it was a virus, but not everyone believed that the doctors knew. One magazine article had said it was related to diet. Another article said it was related to the color of your eyes. Kids at summer camp got it, and when a boy at a camp in upstate New York got it in the summer of 1953, a health officer said no one would be let out of the camp till the polio

season was over. Someone said that public gatherings had been banned altogether in the Yukon. In Montgomery, Alabama, that summer the whole city broke out; more than eighty-five people caught it. An emergency was declared, and in Tampa, Florida, a twenty-month-old boy named Gregory died of it. Five days later, his eight-year-old sister, Sandra, died of it while their mother was in the delivery room giving birth to a new baby.

5 The newspaper published statistics every week. As of the Fourth of July, newspapers said there were 4,680 cases in 1953—more than there had been to that date in 1952, reckoned to be the worst epidemic year in medical history, in which the final tally had been 57,628 cases. But none of the numbers were reliable; odd illnesses were added to the total, and mild cases went unreported. Nonetheless, the totals were not the most terrifying thing about polio. What was terrifying was that, like any plague, you never knew where or when it might strike. It was more random than roulette—only it did seem to strike children disproportionately, and so it was called infantile paralysis—and it made parents crazy with anguish.

6 The rules were: Don't play with new friends, stick with your old friends whose germs you already have; stay away from crowded beaches and pools, especially in August; wash hands before eating; never use another person's eating utensils or toothbrush or drink out of the same Coke bottle or glass; don't bite another person's hands or fingers while playing or (for small children) put another child's toys in your mouth; don't pick up anything from the ground, especially around a beach or pool; don't have any tooth extractions during the summer; don't get overtired or strained; if you get a headache, tell your mother.

7 Nevertheless, kids caught it. In the big city hospitals, kids were stacked like cordwood in the corridors. Carts and wheelchairs congested the aisles. The dominant odor was of disinfectant. The dominant taste was of alcohol-disinfected thermometers. In the Catholic hospitals, holy medals and scapulars' covered the motionless arms and hands of the children. On the South Side of Chicago, a mother cried just to see the name above the door of the place her child was taken: the Home for Destitute Crippled Children. In some places, parent's were allowed to visit their children only once a week—not because of any special fact about polio, only because that was how children's wards were run in 1953. A child in bed with polio never forgot the sound made in the corridor by his mother's high-heeled shoes.

8 Injections of gamma globulin were prescribed for those who had not yet caught it. Certain insurance against measles, gamma globulin did not prevent catching polio, but it did seem to minimize the crippling effects. It was in short supply. Injections were given only to pregnant women and those under the age of thirty who had had a case of polio in the immediate family—or to prevent the spread of an epidemic. The precious supplies were placed under the administration of the incorruptible Office of Defense Mobilization.

9 In Illinois, rumors spread of bootleg gamma globulin. If you were lucky enough to qualify for a shot, you had to endure the humiliation that went with it: you had to pull down your pants and say which buttock would take the inch-long needle. To buy off your pride, the doctor gave you a free lollipop.

10 When the epidemic broke out in Montgomery, Alabama, the story was that 620 volunteer doctors, nurses, housewives, and military personnel administered sixty-seven gallons of gamma globulin (worth $625,000), thirty-three thousand inch-long needles, and thirty-three thousand lollipops. In New York, parents picketed the health department for twenty-seven hours to get it for their children. In some places people said that parents were bribing local officials for vials of gamma globulin. At the same time, an article in the June issue of *Scientific American* reported there was doubt that the stuff was worth a damn. The *New York Times* reported that one little girl came down with polio within forty-eight hours of getting a gamma globulin shot.

11 In the hospitals, meanwhile, children—shrouded in white gowns and white sheets, nursed by women in white surgical masks, white dresses starched to the smooth brittleness of communion wafers—lay in dreadful silence, listening to the faint whispers of medical conversations on the far side of drawn white curtains, the quiet shush of soft-soled nurses' shoes, and the ever-present sound of water in a basin, the ceaseless washing of hands.

12 Parents stood at a distance—six feet from the bed—wearing white gowns and white masks.

13 One boy's uncle gave him a plastic Hopalong Cassidy bank when he was in the isolation ward. After the customary two-to-three week stay there, after the fevers passed, he was moved into the regular children's ward. On the way, the nurses discarded the contaminated bank along with its savings.

14 Some children were not told what they had (lest it be too dangerous a shock to them), and so they discovered for themselves. One boy acquired from his visitors the biggest collection of comic books he had ever had. When he dropped one, he jumped out of bed to pick it up, crumpled in a heap and found he couldn't get up off the floor again.

15 Some would recover almost entirely. Some would die. Some would come through unable to move their legs, or unable to move arms and legs; some could move nothing but an arm, or nothing but a few fingers and their eyes. Some would leave the hospital with a cane, some with crutches, crutches and steel leg braces, or in wheelchairs—white-faced, shrunken, with frightened eyes, light blankets over their legs. Some would remain in an iron lung—a great, eighteen-hundred-pound, casketlike contraption, like the one in which the woman in the magic show (her head and feet sticking out of either end) is sawed in half. The iron lung hissed and sighed rhythmically, performing artificial respiration by way of air pressure.

16 Some moaned. Some cried. Some nurtured cynicism. Some grew detached. Some were swept away by ungovernable cheerfulness. Rarely did anyone scream in rage, however common the feeling. All were overpowered, all were taught respect—for the unseen powers of nature, the smallness of human aspiration, the capacity for sudden and irrevocable change, the potential of chance.

17 As it happened, in the spring of 1953, Dr. Jonas Salk, an insignificant-seeming fellow with big ears, a receding hairline, and a pale complexion, had published a paper in a scholarly journal, reporting that he had induced the formation of antibodies against three types of polio viruses. He hadn't quite fully tested it, he hastened to say, but he had tried it on 161 children and adults with no ill effects. When newspapers got hold of the story, parents phoned their family doctors. Those with medical connections tried to find a way to get to Salk. Salk became famous in an instant—and from the moment of his first announcement, such an outpouring of hope and gratitude attached to him that he came to stand, at once, as the doctor-benefactor of our times.

18 During the summer of 1953, reporters called him weekly for news of progress. His vaccine, he explained, was a dead-virus vaccine. He devastated the virus with formaldehyde and then whipped it up into an emulsion with mineral oil to fortify it, and in this way he thought he had something that, when it was injected into a person,

would stimulate a person's natural defense mechanisms to produce antibodies. However, he was not able to hurry the testing process along. In May 1953, he expanded the test to include more than seven hundred children. And not until the spring of 1954 were more than a million children inoculated in a large field trial financed by the March of Dimes, and, as the papers said, the "total conquest of polio" was in sight. Within the next half dozen years, the Salk vaccine reduced the incidence of polio by perhaps 95 percent, preventing maybe as many as three hundred thousand cases of polio in the United States.

19 Yet Salk's triumph did not last for long. The March of Dimes, in its own need for publicity and contributions, lionized Salk mercilessly—and his fellow doctors soon got tired of his fame. He was not—and never has been since—invited to join the National Academy of Sciences. And soon enough, Salk's colleagues began to point out that Salk, after all, had made no basic scientific discovery. Many people had been working on a preventive for polio. The basic discovery had been made by three fellows at Harvard—Doctors Enders, Weller, and Robbins—who had shown that a polio virus could be grown in certain tissue cultures of primate cells. Before the Harvard finding no one had been able to make a vaccine because no one had been able to cultivate the virus in test-tube cultures. After the Harvard finding, Salk's vaccine was mere applied science. (The Harvard doctors got the Nobel; Salk did not.) Salk had just pulled together the work of others. And some of the others thought Salk had been premature in publishing his paper, that he was rushing his vaccine into the world incautiously. Then, in 1955, a batch with live virus slipped out, and 260 children came down with polio from having taken the Salk vaccine or having contact with persons who had taken it.

20 Meanwhile, even as Salk's vaccine was eliminating polio in the United States, it was already obsolescent. Dr. Albert Sabin, a researcher who told interviewers that work was his recreation, was coming up with a new vaccine. His vaccine used an attenuated (that is to say, live) virus with special properties to stimulate the production of antibodies, and it seemed to offer immunity for much longer than Salk's vaccine, possibly for many years. This virus retained the capacity to multiply in the intestinal tract, thus passing from someone who had received the vaccine to someone who hadn't and inoculating them as well. The Sabin vaccine could be stored indefinitely in deepfreeze units; it could be taken orally and produced cheaply. It was given ex-

tensive tests in 1958 and 1959. By 1962 it had replaced the Salk vaccine almost entirely in the United States and most of the rest of the industrialized world. Although Sabin never got the Nobel either, in the next two decades his vaccine prevented perhaps two to three million cases of polio.

21 But Sabin's happiness was not uncomplicated, either. Though no one likes to mention it—and it does not diminish the good of the vaccine, since the odds are only "one in six or seven million"—sometimes a Sabin inoculation would be, as one specialist in polio has said, "associated with" a case of polio: the attenuated vaccine can never be as absolutely safe as the dead-virus vaccine.

22 Moreover, while the Sabin vaccine has eliminated polio in most of the temperate-climate countries where it's been used, it has not done so well elsewhere: in the Third World, it turns out, polio has not been ended at all. There, uncertain conditions of refrigeration cause the Sabin vaccine to break down. For some reason, too—perhaps because people in parts of the Third World carry other viruses in their systems that interfere with the polio vaccine—some inoculations don't take. The Sabin vaccine does not work with just one dose but requires several doses, which involves massive vaccination of a community. This has been accomplished in Cuba and Brazil but the logistics are staggering in many Third World countries. Despite the inoculation programs of the past two decades, about 375,000 people come down with polio every year in the Third World: seven and a half million in the last twenty years.

23 Some highly refined ironies: At the moment, conditions of sanitation and hygiene are so bad in the Third World that many children come down with polio before the age of two. Fortunately, however, at that age polio comes and goes often without leaving a trace of paralysis. As physical standards of living improve, children will not get polio at such early ages: they will get it instead when they are teenagers, when the paralytic rate is higher. So as health conditions improve in the Third World polio may well increase, increasing the need for vaccination.

24 Some say now that the Salk vaccine will make a comeback, that it will work where the Sabin vaccine has not worked—that the Salk vaccine will hold up better under the conditions of Third World refrigeration, that there is even some indication that a more potent Salk-type vaccine might require only one or two inoculations. Recent tests by the

Israeli government in the Gaza Strip seem to make a case for the Salk vaccine. A French pharmaceutical company is manufacturing a Salk-type vaccine that also vaccinates against diphtheria, tetanus, and whooping cough. It may be that Salk will become famous again.

25 These days, as polio continues to occur in the Third World, most of those who gather at the special conferences on the disease feel that the old Salk vaccine—which has continued to be used in some of the smaller European countries—ought to be brought back on a large scale. Most of them feel not that the Salk should replace the Sabin but rather, given everyone's doubts, that both vaccines are needed, in different circumstances, or perhaps in combination.

26 But when the two grand old men of the fight against polio, Salk and Sabin themselves appear at these conferences, they disagree. Each man—as modest and thoughtful and impressive as he is in private—takes on a missionary zeal in public, strutting and scrapping for pre-eminence, each arguing for the ultimate superiority of his own vaccine. Sabin argues politics: the administration of his vaccine must be improved. Salk argues effectiveness is possible with fewer doses with his vaccine and warns of live-virus-vaccine-associated polio. At one such recent encounter, Salk tried everything, even charm and banter, to win over the audience; he and Sabin agreed on only one thing, he said with a skilled debater's smile, "that only one vaccine is necessary."

27 And so the two renowned old doctors go on grappling with each other and with themselves, speaking not only of the progress of science and the triumph of reason but also—like those of us who got polio in the summer of 1953 and have toted around a couple of canes ever since—of the equivocalness of greatness, the elusiveness of justice, the complexity of success, the persistence of chance.

Suggestions for Discussion and Writing

1. Where does the introduction to this essay end? Why does Mee choose to introduce the story of Salk and Sabin in this manner? What effects does he induce in his audience by doing so?

2. Do you suspect that Mee contracted polio as a child? What clues in the essay lead you to your conclusion?

3. Why is the battle between Sabin and Salk important to understanding the polio vaccine? What are the advantages and disadvantages of each type of vaccine?

4. Why did the Harvard researchers get the Nobel Prize for the polio vaccine instead of Salk or Sabin? What point is Mee making about the scientific community by pointing this out?

5. Most communities require vaccinations for communicable diseases; traditionally students must have them before enrolling in classes for the first time. Sometimes, however, children experience serious side effects from receiving these vaccinations; these side effects can't be predicted in advance. What is the rule on vaccinations in your community? What would happen if parents refused to have their child vaccinated? Whose rights should prevail—the individual's or the community's? Assume you are the assistant to the legislator in your community. What position would you advise her to take on this question?

LeRoy Walters

Ethical Issues in the Prevention and Treatment of HIV Infection and AIDS

LEROY WALTERS was born in Sterling, Illinois, in 1940. After taking his undergraduate degree and graduating from the seminary, he studied in Germany, before receiving his master's and doctoral degrees in philosophy from Yale. He is a Fellow and former director of the Center for Bioethics, Kennedy Institute of Ethics at Georgetown University; he serves on the boards of many medical and religious ethics foundations. With Tom L. Beauchamp, he is the editor of *Contemporary Issues in Bioethics* (1978, 1982, and

1989), and he is the editor of the annual *Bibliography of Bioethics.* Recently he has written about the ethics of human gene therapy and about new reproductive technologies. This essay first appeared in *Science* in 1988.

An adequate ethical framework for evaluating public policies regarding infection with the human immunodeficieny virus (HIV) will include the following considerations: (i) the outcomes, often categorized as benefits and harms, of the policies; (ii) the distribution of these outcomes within the population; and (iii) the liberty-rights, or freedoms, of those who are affected by the policies. A recent presidential commission on bioethics called these three considerations well-being, equity, and respect. In their *Principles of Biomedical Ethics,* Beauchamp and Childress designate these considerations beneficence and nonmaleficence, justice, and respect for autonomy.

2 As we and other societies attempt to confront the AIDS epidemic, the central problem we face is the following: How can we control the epidemic and the harm that it causes without unjustly discriminating against particular social groups and without unnecessarily infringing on the freedom of individuals? This formulation accepts the importance of halting the transmission of HIV infection but recognizes that the achievement of that goal may at times be in tension with other moral constraints, namely, constraints based on justice or respect for autonomy. At the same time, however, these three considerations, or moral vectors, may all point in the same direction, for example, if a particular policy is simultaneously counterproductive, discriminatory, and intrusive.

3 In this article I will indicate how the ethical principles of beneficence, justice, and respect for autonomy relate to the epidemic of HIV infection in the United States. I will argue that, because these three principles are all of importance, none of them should be ignored in the formulation of public policy. While one principle may predominate in a given situation or sphere, it should not be allowed to overwhelm or displace the other two.

4 Three types of policies will be considered: public health policies, policies for the delivery of health care to people with HIV infection, and research policies.

Public Health Policies

5 **Public education.** Until more effective medical therapies and preventive measures are developed, public education is likely to be one of the most important means for controlling the epidemic. If the education appeals to the rational capacities of the hearer, it respects his or her autonomy. If public education simultaneously leads to risk-reducing behavioral change, it also promotes the health of the hearer and his or her associates.

6 Imaginative public education will be moral education in the sense that it helps the hearer to see clearly the possible effects of his or her behavior on others. One possible approach to such education involves the use of ethical if-then statements such as the following. "We have discussed the pros and cons of engaging in behavior X. If you choose to do X, then, in order to avoid harming others, you should adopt measures A, B, and C." Fortunately, many of the measures that protect others are also self-protective. Thus, public educators can simultaneously appeal to both the self-interested and altruistic sentiments of their audiences.

7 While everyone who is at risk of contracting or transmitting HIV infection should be educated, there are strong moral arguments for targeting educational efforts especially toward people who are most likely to engage in risky behaviors—for example, receptive anal intercourse, intravenous (IV) drug use with shared needles, or vaginal intercourse with IV drug users. Such targeted programs can be justified on either or both of two grounds. Intensive coverage of the groups most at risk for infection is likely to be more efficient in controlling the epidemic than general educational programs alone will be. It can also be argued that groups at higher than average risk need, or even deserve, stronger than average warnings of the risks to which they may be exposed.

8 **Modified approaches to IV drug use.** Twenty-five percent of clinical AIDS cases involve the illegal use of IV drugs. The sharing of needles and syringes, sometimes a ritual in settings where multiple drug users self-inject together, seems to be the principal mode of transmission among IV drug users. People who become infected through sharing contaminated needles and syringes may, in turn, infect nondrug-using people through sexual intercourse.

9 Members of ethnic and racial minority groups are represented in disproportionate numbers among U.S. IV drug users who have

AIDS. In AIDS cases involving IV drug use as the sole risk factor, 51% of patients are black and 28% are Hispanic. Among minority group women, the correlation between IV drugs and AIDS is particularly strong: 70% of black women with AIDS and 83% of Hispanic women with AIDS are either IV drug users or the sexual partners of IV drug users. Two-thirds of black children and three-fourths of Hispanic children with AIDS contracted their infections from mothers who were members of the same two risk groups.

10 It is clear that current programs for IV drug users in the United States are failing in many respects and that new and bold measures are needed. These measures may not be politically popular, given the misunderstanding and fear that frequently surround drug use and given our society's traditional neglect of IV drug users. But the initiatives will be essential for controlling the epidemic, for meeting the needs of people who are often stigmatized, and for enabling IV drug users to make autonomous choices about their lives.

11 The first initiative that should be undertaken is the expansion of drug-treatment programs to accommodate, on a timely basis, all IV drug users who desire treatment. Reports of 3-month waiting lists in U.S. drug-treatment programs are commonplace. Our failure to provide treatment to people who indicate an interest in discontinuing drug use is both short-sighted and counterproductive. It is encouraging to note that the Presidential Commission on the Human Immunodeficiency Virus Epidemic is making the lack of programs to treat IV drug users one of four major areas for initial study.

12 A second initiative that will probably be necessary to control the epidemic among IV drug users is the establishment of public programs for the exchange of sterile needles and syringes for used and possibly contaminated equipment. Three countries, the Netherlands, the United Kingdom, and Australia, have experimented with free needle-exchange programs and have reported initially encouraging results—although it is too early to know for certain that the exchange programs actually reduce the rate of infection transmission. Proposals to initiate needle-exchange programs in the United States have not yet been implemented, in part because they appear to condone or even to encourage IV drug use. Perhaps for this reason U.S. law-enforcement officials have generally opposed such programs. However, the ethical if-then statements discussed above may also pertain here. Moral and legal prohibitions of IV drug use have not achieved universal acceptance in

our society. Given that fact, one seeks to formulate rules of morally responsible drug use: "If you choose to use IV drugs, then you should take steps, including the use of sterile needles and syringes, to minimize the chance of your becoming infected and infecting others with HIV."

13 If the foregoing measures, coupled with targeted education for IV drug users, are insufficient, more radical initiatives will need to be contemplated. One of the most controversial initiatives, at least among law-enforcement officials, would be the provision of controlled access to injectable drugs by IV drug users in an effort to bring addiction and its social context above ground. Such a policy was endorsed editorially in May 1987 by the British journal *The Lancet.* Pilot programs of controlled access to injectable drugs, with simultaneous decriminalization of IV drug use, could provide valuable data on the potential effectiveness of this initiative.

14 **Modified approaches to prostitution.** Male or female prostitutes who have unprotected intercourse with multiple sexual partners expose themselves to considerable risk of HIV infection in areas of moderate to high seroprevalence. This theoretical risk has been actualized among female prostitutes who have been studied epidemiologically in both the United States and equatorial Africa. For example, a recent cross-sectional survey of female prostitutes in the Newark, New Jersey, area indicated that 51.7% tested positive for antibody to HIV in 1987; in Miami the seroprevalence rate among incarcerated female prostitutes was 18.7%. A high infection rate among prostitutes imperils not only their own health but also the health of their clients and their clients' other sexual partners.

15 Official policies on prostitution in this country are set by states and localities. In most U.S. jurisdictions the general approach has been to criminalize the practice of prostitution; in some jurisdictions, the patronizing of a prostitute is also a crime. In contrast, many European countries and several counties in Nevada have adopted a licensing or regulatory approach that includes periodic screening of prostitutes for infectious disease.

16 An ethically appropriate response to prostitution will be based not simply on our evaluation of prostitution as a practice but also on careful assessment of the extent to which alternative public policies on prostitution are compatible with the principles of beneficence, respect for autonomy, and justice. Although the intervening variables are numerous, the available evidence from Nevada and Europe suggests

that, compared with the outlaw and arrest approach, the licensing and regulatory approach to prostitution is at least correlated with lower rates of infection with several other sexually transmitted diseases among prostitutes. At the same time, a licensing and regulatory approach displays greater respect for the autonomy of adult persons to perform acts that affect chiefly the persons themselves, especially if the transmission of disease is prevented through the use of condoms and through regular health examinations.

17 Again in this case we should be willing to become pragmatic and experimental in our approach to controlling the epidemic. The legal prohibition of prostitution has not been notably successful in preventing a rapid rise in seropositivity among prostitutes, at least in some cities. Pilot studies of less restrictive approaches in selected localities, taken together with evidence from Nevada and Western Europe, might reveal that alternative policies are, on balance, ethically preferable.

18 **Modified approaches to homosexual and bisexual sexual activity.** As of December 1987, 65% of AIDS cases in the United States involve homosexual or bisexual males; an additional 8% of cases involve homosexual or bisexual males who also admit to IV drug use. Thus, in a substantial fraction of U.S. AIDS cases to date, HIV seems to have been transmitted through sexual intercourse between males. Receptive anal intercourse is one of the principal modes of viral transmission.

19 Many homosexual and bisexual males with AIDS or HIV infection became infected before AIDS was described as a clinical syndrome and before the primary modes of transmission were identified. Thus, while they may have known that they were at increased risk for a series of treatable sexually transmitted diseases, for example, gonorrhea or hepatitis B, they could not have known that they were at risk for contracting an infection that might lead to AIDS. Since the facts about HIV transmission have become well known, homosexual and bisexual men have been heavily involved in targeted public education programs and in humane health care programs for people with AIDS. There is also substantial evidence to indicate that considerable numbers of homosexual and bisexual males have altered their sexual practices to reduce their probability of becoming infected and infecting others with HIV.

20 It might seem that, short of traditional public health measures

such as increased testing and screening, little more can be done to encourage the cooperation of homosexual and bisexual males in controlling the epidemic. However, two public policy initiatives might conceivably have a salutary effect: (i) in jurisdictions that currently outlaw such acts, the decriminalization of private homosexual acts between consenting adults; and (ii) in jurisdictions that currently lack such antidiscrimination statutes, the legal prohibition of discrimination against people who engage in private consensual homosexual acts.

21 There would be strong moral arguments for these legal changes even in the absence of a major epidemic. However, in the midst of an epidemic that has already affected large numbers of homosexual and bisexual men, the following additional arguments can be advanced. First, decriminalization and antidiscrimination initiatives would encourage homosexual and bisexual males to disclose their patterns of sexual activity to health providers and hospitals without fear that a breach of confidentiality could lead to criminal prosecution. Such open disclosure could, in turn, lead to the discussion of risk-reducing practices such as the use of condoms or the avoidance of anal intercourse. Second, the legal changes could facilitate the gathering of more accurate data on current patterns of sexual activity in the United States—patterns that have not been studied in depth since the research of Alfred Kinsey and associates in the 1940s. By reducing respondents' fears about being stigmatized, the proposed legal changes could enhance the accuracy of data that could then be used for educational and epidemiological purposes. Third, the proposed legal changes would send a clear signal to homosexuals and bisexuals that heterosexuals intend to treat them with what Ronald Dworkin has termed "equal concern and respect." More specifically, these policies would help all of us, regardless of sexual orientation or pattern of sexual practice, jointly to reassess whether the magnitude of our national effort to control the current epidemic has been proportionate to the gravity of the threat posed by the epidemic.

22 **Testing and screening programs.** The moral and legal justification for testing individuals or screening populations for antibody to HIV has been extensively debated. James Childress has proposed a helpful taxonomy of screening programs:

Extent of screening	Degree of voluntariness	
	Voluntary	Compulsory
Universal	1	2
Selective	3	4

A recent amendment to the voluntary category in this matrix is "routine" counseling and testing, which is defined in Public Health Service guidelines as "a policy to provide these services to all clients after informing them that testing will be done." The Public Health Service guidelines add that, "Except where testing is required by law, individuals have the right to decline to be tested without being denied health care or other services."

23 There is scant justification and little public support for universal HIV antibody screening programs, whether voluntary or compulsory. The principal arguments against such programs are consequential. The usual screening test has poor predictive value in populations where the prevalence of seropositivity is low; thus, large numbers of people who are in fact antibody-negative would be falsely identified as positive during initial screening. Further, the cost of universal screening would be high, especially given the fact that screening would need to be repeated at regular intervals to track changes in antibody status. In short, universal screening is incompatible with the principle of beneficence. Mandatory universal screening would involve a massive violation of the respect for autonomy principle, as well.

24 The crux of the current debate is whether selective screening for HIV antibody should be undertaken and, if so, whether the screening should be compulsory or voluntary. To date, most commentators on the ethics of HIV antibody screening have argued that only carefully targeted, voluntary screening programs are morally justifiable and that such programs are morally justified only if they fulfill three conditions: (i) the programs include adequate counseling of screenees; (ii) they protect the confidentiality of information about individuals, except in carefully specified exceptional circumstances; and (iii) they are conducted in a context that provides guarantees of nondiscrimination to seropositive individuals. Categories of persons often nominated for selective, voluntary screening programs include hemophiliacs, IV drug users, homosexual and bisexual men, prostitutes, patients

at clinics for sexually transmitted diseases, heterosexual sexual partners of infected persons, prisoners, military recruits and personnel, applicants for marriage licenses, and hospital patients, especially patients undergoing surgery or hemodialysis.

25 It is not possible here to discuss each of these population groups. I will, however, comment on the three conditions for ethically acceptable voluntary screening programs. The provision of face-to-face counseling to all persons participating in a large-scale, voluntary screening program may be infeasible on financial grounds. Thus, it might at first glance seem reasonable to reserve counseling for screenees who are confirmed to be HIV antibody positive. However, if voluntary screening programs are targeted to selected groups with much higher than average prevalence, then the screening context would seem an ideal setting for carefully tailored education regarding risk reduction. Such counseling demonstrates a program's respect for the autonomy of screenees and should help to slow the progress of the epidemic, as well.

26 The protection of patient confidentiality in all but carefully delineated circumstances also demonstrates respect for the autonomy of screenees. Guarantees of confidentiality can be strengthened by statutes that impose criminal sanctions for unauthorized, medically nonindicated disclosure of antibody status. At the same time, however, guarantees of confidentiality should not be absolute. Several commentators have argued, for example, that health care providers have a moral duty to warn known intimate associates of an antibody-positive person who refuses to inform the associates of his or her antibody status and who continues to place those associates at risk. In this case, the health care provider cannot simultaneously respect the autonomy of both the screenee and the associates.

27 The level of participation in voluntary screening programs is likely to be higher if legal guarantees against discrimination are provided to antibody-positive persons. These guarantees would complement the general guarantees of nondiscrimination discussed above. One formulation of such a guarantee in a major federal bill reads as follows:

> A person may not discriminate against an otherwise qualified individual in employment, housing, public accommodations, or governmental services solely by reason of the fact that such individual is, or is regarded

as being, infected with the etiologic agent for acquired immune deficiency syndrome.

28 In a democratic society, the presumption should be in favor of voluntary rather than mandatory public health programs. This presumption should be overridden only as a last resort, after voluntary alternatives have been vigorously employed and have failed, and only if there is a reasonable hope that a mandatory program would succeed. In my judgment, voluntary screening programs that include adequate counseling and appropriate guarantees of confidentiality and nondiscrimination have not yet received a sufficient trial in the United States. Such screening programs, coupled with anonymous testing for those who desire it and with the other public health strategies outlined above, offer us a reasonable hope of bringing the AIDS epidemic under control. Thus, I conclude that mandatory screening programs—other than those involving persons who voluntarily donate blood, semen, or organs—are not morally justifiable at this time.

Policies for the Delivery of Health Care

29 Even as public health effort to prevent the further spread of HIV infection proceed, some of the approximately 1.5 million already-infected people in the United States will experience initial symptoms, become ill, develop full-blown AIDS, or die. As of 7 December 1987, 47,436 infected adults and 703 infected children had been diagnosed as having clinical AIDS; 26,816 (60.3%) of the adults and 419 (59.6%) of the children had died. HIV infection produces a broad clinical spectrum that includes, at its extremes, asymptomatic status and terminal illness. The health care delivery issue is currently focused on people who are symptomatic as a result of HIV infection and who know that they are infected with HIV. Increasingly, however, people at risk for HIV infection are likely to call on the health care system for help in clarifying their antibody status. Further, the health care system may be able to offer medical interventions to asymptomatic infected people that will prevent, or at least delay, some of the possible sequelae of HIV infection.

30 **The duty to provide care.** This issue can be considered at two levels: the level of the individual health care worker and the level of health-related institutions and the health care system.

31 Surveys of attitudes toward caring for AIDS patients in one high-prevalence area have revealed considerable anxiety among physi-

cians and nurses. A study conducted at four New York residency programs in 1986 noted that 36% of medical house officers and 17% of pediatric house officers reported needlestick exposure to the blood of AIDS patients. Twenty-five percent of respondents indicated that they "would not continue to care for AIDS patients if given a choice." A 1984 survey of nurses at the Westchester (New York) County Hospital found that 39% would ask for a transfer if they had to care for AIDS patients on a regular basis.

32 Studies suggest that the probability of infection transmission from patient to health care worker is very low. Yet ten reasonably well-documented cases of seroconversion in health care workers have been reported, with six of these workers having been exposed by accidental needlesticks and the remainder by exposure of the eyes, mouth, or hands and arms to infectious body fluids. HIV seems to be much less infectious than the hepatitis B virus. Yet this comparison is not entirely pertinent; hepatitis B is not usually a lethal disease, and an effective vaccine against the disease is available. Thus, there remains a very small but nonetheless real probability that health care workers will acquire HIV infection from the blood or other body fluids of people with HIV infection. In an unknown proportion of these workers, the infection will have lethal consequences.

33 Despite these attitudes and risks, it might seem at first blush that the ethical obligation of health care workers to care for people with HIV infection is clear. The words "profession" and "professional" leap readily to mind, as do images of real or fictional heroines and heroes such as Florence Nightingale, Benjamin Rush, or Bernard Rieux. Yet the scope of the term health care worker is broad and includes the medical technologist, the phlebotomist, and the person who transports infective waste to the incinerator. Further, the basis for and the extent of the health care worker's obligation to provide care for patients are matters of dispute—despite several vigorous reassertions of the physician's moral duty to treat people with HIV infection.

34 A reasonable ethic for health care workers will not require of them heroic self-sacrifice or works of supererogation. Such a requirement would violate both the principles of autonomy and beneficence. On the other hand, a reasonable ethic will not allow people who are in need of care to be refused treatment or abandoned solely because they are infectious. Such refusal and abandonment would violate the principle of beneficence. Universal infection-control precautions

such as those suggested by the Centers for Disease Control (CDC) are likely to reduce substantially the risks to health care workers; thus, heroic self-sacrifice will not be required. If these measures are insufficient in certain high-risk settings, or if the universal precautions seriously impede patient care, testing of selected categories of patients, for example, surgical patients, may be justifiable. This testing should be carried out only with the prior knowledge and consent of patients and should include counseling for seropositive persons. Patients who decline testing will be presumed to be antibody-positive. Testing measures will seem less threatening to patients when carried out in a social context that respects confidentiality and opposes discrimination.

35 At the level of health care institutions and the health care system, the AIDS epidemic has exacerbated already existing problems regarding access to health care. The access problems faced by people with AIDS or HIV infection do not differ qualitatively from those faced by many other U.S. citizens with chronic or terminal illness. However, because people with HIV infection are almost always under 65 years of age, their health care needs graphically illustrate major deficiencies in the current U.S. system for providing health care to the nonelderly.

36 Even before the AIDS epidemic became a major factor in health care financing, it was almost commonplace to assert that 15 to 17.5% of U.S. residents under age 65 lack both public and private health insurance. These percentages translate into 30 to 35 million Americans. An additional 10 to 15% of these under age 65 who are insured are not adequately protected against chronic or catastrophic illness. Of the 150 million Americans under 65 who are privately insured, at least 80% have their health insurance tied to group plans at their place of employment.

37 People with HIV infection who are currently employed and who have group health insurance coverage through their employers are in the best position to cope with the medical costs that may result from their infection. However, even for these most well-off people a double threat looms. If they become so ill that they can no longer continue employment, they face the prospect of losing both their source of income and their group health insurance coverage. Although federal legislation enacted in 1985 provides for continuing individual health insurance coverage for 18 months after the termination of em-

ployment, the cost of such coverage may be prohibitive for an unemployed person. Other people with HIV infection who become symptomatic—the underinsured, the uninsured, and the unemployed—are likely to rely on Medicaid for assistance, if they can meet complex eligibility requirements. Actuaries from the Health Care Financing Administration estimate that 40% of patients with clinical AIDS are assisted by Medicaid with their direct medical care expenses and that an average of 23% of such expenses are borne by Medicaid. In fiscal year 1987 federal and state Medicaid expenditures for AIDS patients were estimated at $400 million.

38 The future looks bleak, both in terms of costs and in terms of shortages of needed services for chronically and terminally ill patients. In the cost projections made to date, the estimates of personal medical costs for AIDS patients alone in 1991 range from a low of $3.5 billion to a high of $9.4 billion (in 1984 dollars). Already in 1988, there are shortages of nursing home facilities, home care programs, hospice facilities, and counseling services for clinically ill people with HIV infection.

39 Divergent views exist about the appropriate role of the private sector in the provision of health care to people infected with HIV, as well as to other people with health care needs. What is clear, however, is that we as a society cannot expect private hospitals and nursing homes to operate at a loss. Nor can we expect private health insurers or self-insuring employers to ignore the financial impact of an unanticipated epidemic.

40 The central ethical question confronting the U.S. health care system was evident long before HIV was discovered or named. That question is: Does our society have a moral obligation to provide some level of health care to every one of its members? Several commentators on the ethics of health care allocation have argued that our society does have such an obligation. They have based their argument on the principles of beneficence (the unpredictability of health care needs and the harms caused by lack of access) and justice (the inequities that result from current differentials in access). They assert that the principle of respect for autonomy must take second place, as those of us who are financially well off are called upon to share in meeting the needs of the less well off, presumably through the payment of increased premiums and taxes.

41 This judgment seems to me to be correct. If so, the major policy

question is no longer whether we should attempt to meet the needs of the medically less well off. Rather, we should address the questions "What constitutes a basic level of care?" and "How can this level best be provided to everyone, including people infected with HIV?"

42 **Neurological involvement and consent to care.** An unknown proportion of people with HIV infection experience involvement of the central nervous system, including the brain. Indeed, the CDC has recently expanded the clinical definition of AIDS to include such neurological complications. The extent of neurological involvement may range from minor symptoms of cognitive impairment to totally disabling dementia.

43 Brain involvement resulting from HIV infection, like brain involvement due to other causes, inevitably complicates the relation between patient and health provider. Two methods of extending patient autonomy forward in time have proved helpful in other health care settings and may also be beneficial in the treatment of HIV-infected patient with early symptoms of neurological deterioration. Advance directives about preferred modes of care or nontreatment are now expressly recognized by the statutes of 38 states and the District of Columbia. In addition, 18 states make legal provision for a patient's appointment of a spokesperson with durable power of attorney, who can express the patient's wishes if the patient should become incapacitated or be adjudged legally incompetent. The patient's spokesperson is usually a trusted friend or family member. Both modes of anticipatory decision-making were strongly endorsed by the President's Commission on Bioethics in 1983 and both seem well adapted to the needs of HIV-infected patients with neurological symptoms.

44 **The care of dying AIDS patients.** When treatment fails and death within a few months becomes inevitable, people with AIDS deserve compassion and support. Individual patient preferences vary, but many terminally ill patients have expressed a desire to die at home in the company of friends or in a hospice-like institutional setting. These alternatives should be provided by an upgraded system of care for all terminally ill patients.

45 A central role in patient management should be played by the patient's own directives and, if the patient becomes mentally incapacitated, by the patient-designated proxy. If at all possible, future decisions about resuscitation and the use of artificial nutrition and hydration measures should be explicitly discussed with the competent AIDS

patient. Like other terminally ill patients who face the probability of severe physical deterioration and the possibility of a painful death, some AIDS patients will also want to discuss the options of suicide or voluntary active euthanasia. Both of these topics have received intensive study, especially in the Netherlands, the United Kingdom, and the United States. Respect for the autonomy of terminally ill patients would seem to require us to place these difficult issues on the agenda for sustained local and national discussion.

Research Policies

46 In the long term, the best hope for controlling the AIDS epidemic lies in biomedical research. A vaccine against HIV would seem to be the ideal solution but if immunization strategies prove to be infeasible, chemoprophylactic measures may succeed. For people already infected with HIV, new interventions are under development, but progress has been slow. Epidemiological, social-scientific, educational, and social-intervention studies will also be key elements in an overall research strategy.

47 A general question that has been raised about the U.S. research effort is whether it has been proportionate to the gravity of the threat posed by the current epidemic. A 1986 report from the Institute of Medicine and the National Academy of Sciences concluded that at that time the response was inadequate. A less than adequate response to the epidemic violates both the principle of beneficence and the principle of justice. It fails to prevent avoidable harm to thousands if not millions of people, and it conveys the impression that policy-makers do not care about the welfare of the groups most at risk. Even in the best of times, members of several groups at increased risk for HIV infection experience neglect or even stigmatization by many of their fellow citizens. These are not the best of times.

48 Clinical trials of various treatments are being conducted in asymptomatic and symptomatic people with HIV infection as well as in patients with clinical AIDS. The usual practice in early trials is to use a placebo-controlled design with each subgroup of people until an effective therapy for that group is discovered. When the efficacy of an agent has been demonstrated, placebos are no longer given; rather, various dosages of the effective agent are compared, or a new candidate therapy is compared against the older, effective therapy.

49 Some critics have questioned whether it is ethical to conduct

placebo-controlled trials with HIV-infected patients. Some have suggested that all symptomatic people with HIV infection should be given immediate access to potentially promising therapies that have not been validated in randomized controlled trials. Here one can, in my view, make a justice-based argument for subjecting potential treatments for HIV infection to the same kind of rigorous study that other new treatments must undergo. Further, from the perspective of beneficence, unnecessary suffering would be visited on people with HIV infection if they were provided immediate access to ineffective "therapies" or treatments with toxic effects that far outweigh their therapeutic benefits.

50 The testing of vaccines for the prevention of HIV infection will also raise important ethical questions. For example, it will be necessary for uninfected volunteers to be exposed to inoculations that will make them antibody-positive by ELISA and Western blot tests. Further, research subjects who participate in unsuccessful vaccine trials may thereby be made more susceptible to HIV or other infections than they would have been had they not taken part in the trials. Equally disturbing is the possibility that some subjects, having received an ineffective vaccine, may be rendered incapable of being immunized by subsequently developed effective vaccines. Because the numbers of participants in early trials may reach into the thousands or tens of thousands, they could constitute a serious additional public health problem for society.

51 The risks associated with vaccine trials have prompted some researchers to consider testing vaccines against HIV in countries of equatorial Africa, where the prevalence of infection is known to be higher than in the United States and where the number of trial participants could therefore be lower. In addition, the risk of litigation for research-related injury might be reduced in a non-U.S. setting. However, the proposal to export research risks raises questions of fairness in its own right.

52 Partial solutions to the ethical quandaries presented by vaccine trials can be found in policies that exemplify the principles of respect for autonomy, beneficence, and justice. The autonomy of participants in vaccine trials will be respected if they are warned clearly and in advance of the potential physical and social harms to which they will be exposed. The careful planning and foresight of researchers can also reduce the harms associated with vaccine-induced seropositivity. For

example, in a vaccine trial sponsored by the National Institutes of Health, volunteers will be provided with official documentation certifying that their antibody status had been negative before they participated in a vaccine trial. Nonetheless, if participants in vaccine trials are injured as a result of their participation, they may have a legitimate claim to compensation for disabilities incurred in a publicly declared war on a major disease. Indeed, the principle of justice may require the establishment of a compensation program for research-related injuries.

53 **Other types of research.** Epidemiologic research will provide a scientific basis for policies in public health and health care delivery. Longitudinal studies among members of at-risk groups will help to clarify the natural history of HIV infection and the role of cofactors in the development of clinical symptoms. Homosexual and bisexual men, in particular, have been active participants in published longitudinal studies. Cross-sectional studies of demographic groups—newborn infants, patients in "sentinel" hospitals, and residents in selected metropolitan areas—will facilitate more refined estimates of the number of people infected with HIV. One of the major ethical questions in cross-sectional studies has been whether to retain the identifying links between blood samples and the individuals from whom the samples were taken. Anonymous, unlinked testing without consent seems to be emerging as the method of choice, in part because a recent interview survey indicated a likely refusal rate of about 30% among adult Americans if they were invited to be tested in a national seroprevalence study. The advantages of anonymous epidemiologic studies are that no identifiable subjects are placed at risk and that the research results are not skewed by refusals. The disadvantage is that seropositive individuals cannot be identified, notified, and counseled.

54 Other types of research can also play important roles in understanding and coping with the current epidemic. Social science and behavioral research will help to elucidate such questions as the extent of homosexual sexual activity among U.S. adults—a topic that has not been studied in large, rigorously selected samples since the 1940s. Educational research will assist public health officials and counselors in communicating more effectively about lifesaving alternatives in the most intimate realms of human behavior. Finally, social-intervention research can provide public policy-makers with essential information about the effects of innovative approaches to social prac-

tices such as IV drug use and prostitution.

Conclusion

55 At the beginning of this article I mentioned three ethical principles that are thought to be of central importance in contemporary biomedical ethics: beneficence, justice, and respect for autonomy. These principles have informed the preceding analysis. However, as I have reflected on the complexities of the current epidemic, it has occurred to me that a fourth ethical principle may be required to guide our actions and policies in response to this major threat to the public health. I do not have a precise name for this additional principle, but I will venture to suggest some first approximations: mutuality, solidarity, or community.

Suggestions for Discussion and Writing

1. What groups, according to Walters, are most at risk for the HIV virus? In the years since this essay originally appeared, what other groups have also found themselves at risk? How does Walters recommend that each group be treated? Why is it difficult to muster public sympathy for these groups?

2. Why does Walters see the treatment of AIDS as a matter of competing rights? What rights in particular are in question? What are the ethical issues involved in treating HIV/AIDS?

3. *Science* is a publication most often read by people in the science industry: researchers, businesspeople, government policy makers, and so on. How has Walters adjusted his argument for this specialized audience? What sorts of changes would have to be made (for instance in tone, style, details) if you were rewriting it for a "general" audience?

4. What factors most affect policies for delivering health care to HIV-infected people? What are the moral and ethical issues involved in providing such health care?

5. There is only one documented case of an AIDS-infected medical professional who gave the disease to patients (David Acers, the Florida dentist who infected Kimberly Bergalis and four others). There are about one hundred cases where medical personnel contracted AIDS through exposure to infected patients or their bodily fluids. Do these cases justify mandatory AIDS testing for all medical personnel? for all patients? What would be the costs and benefits of such testing?

Ann E. Weiss

The Rights of Patients

ANN E. WEISS, born in Massachusetts in 1943, received her bachelor's degree from Brown in 1965 and now lives with her family in Maine. She has won many awards for her writings, including two Christopher Awards for children's books, and the National Science Teachers' Association and National Council for Social Studies Book Awards. Some of her recent books are *The Supreme Court* (1987), *Lies, Deception, and Truth* (1988), and *Welfare: Helping Hand or Trap?* (1990). This is from *Bioethics: Dilemmas in Modern Medicine* (1985).

Mary E. Schloendorff entered New York Hospital in January 1908. She was complaining of stomach problems.

2 It was a Dr. Bartlett who first examined the woman. His diagnosis: a fibroid tumor. Dr. Bartlett reported this finding to the patient. On so much, everyone was agreed.

3 Then the stories began to diverge. Dr. Bartlett said that he told Mary Schloendorff that an operation was needed to remove the tumor. She agreed to undergo the procedure, and the tumor was promptly taken out.

4 The patient's story was different. According to her, Dr. Bartlett knew she did not want an operation so he informed her that he needed to examine her again, this time under anesthesia. She agreed to this second examination, and was etherized. When she awoke, however, she discovered that an operation had taken place.

5 Angered, Mary Schloendorff went to court, accusing Dr. Bartlett and New York Hospital of deceiving her and of operating without her permission. By 1914, the case had reached the Supreme Court of the state of New York.

6 After listening to arguments on both sides, the court found in favor of the patient. Justice Benjamin Cardozo summed up the majority opinion. "Every human being of adult years and sound mind," he wrote, "has a right to determine what should be done with his own body; and a surgeon who performs an operation without his patient's consent commits an assault, for which he is liable in damages." Cardozo meant that every patient has the right to decide for him—or herself whether or not to accept treatment. If treatment is given against a patient's will that treatment amounts to an assault—a physical attack—upon the patient.

7 Such an attack is a form of malpractice. Malpractice is professional misconduct by a health care professional, misconduct that the patient feels has been harmful to him or to her. If a doctor operates without the patient's permission, that's malpractice. A surgeon who botches an operation, an internist who prescribes the wrong drug, a doctor who sets a broken leg so that the person ends up with a limp—all could be guilty of malpractice. Each may have to repay the patient, in money, for the injury caused.

8 Naturally, health care professionals do not want their patients suing for malpractice. Therefore they and the administrators of the institutions at which they practice require each patient to sign a consent form before undergoing treatment. This protects the patient because it ensures that no procedure will be carried out without his or her permission. A signed consent form protects the hospital and the HCPs too. It is a sort of agreement in advance that the patient will not sue for malpractice.

9 For years consent forms did their job, and few malpractice suits were brought before American courts. Then, about twenty years ago, that began to change. More and more patients—even those who had signed consent forms—were charging malpractice.

10 Why? Medicine has changed in many ways since the early part of the century. With those developments has come a change in people's attitudes toward medical care and toward the men and women who provide it.

A Changing Medical World

11 Years ago, medicine could not accomplish what it can today. Early in this century doctors could not make the sophisticated diagnoses that are now routine. They lacked modern drugs and the surgical

tools and techniques that are commonplace in the 1980s. But twentieth-century improvements in the quality of medical care have not improved the relations between doctor and patient. Often, they have done the reverse.

12 Early in the century, many doctors and patients knew each other better than most do today. People moved around less, and often stayed with just one physician for thirty or forty years. That one doctor treated them for everything. He was trusted and respected, perhaps even loved, by his patients.

13 It's a rare doctor and patient who have that kind of a relationship today. Most doctors now practice in groups. You may go to Dr. Smith for your regular checkups, but if Dr. Smith is off duty when you get flu, you'll have to see Dr. Brown. Next time you're sick, both Smith and Brown may be off duty. You'll end up seeing Dr. Jones. It is hard to build up a solid relationship with any one doctor under such circumstances.

14 When a person falls ill, complications increase. Serious illness requires a specialist. The patient with heart disease sees a cardiologist; one with cancer, an oncologist; one with a blood disorder, a hemotologist. Specialists are inevitable in today's complex and fast-changing medical world. New discoveries are made so frequently that no single doctor can keep up with them all. Each must specialize in a particular area.

15 Since that area can be extremely narrow, one specialist may not be enough. A patient hospitalized with a major illness may see one specialist in the morning and another, whose expertise is in a slightly different area, in the afternoon. Either or both may call in still other doctors. Soon the patient may be under the care of four or five different people, all of them strangers.

16 That makes it hard for patients to find out how ill they are, or how their conditions might be treated. Each doctor may assume that another specialist, or the patient's regular doctor, has clarified all the details. Or each may give the patient a slightly different explanation of what is going on. Either way, patients often feel confused and bewildered. Result: when they sign a consent form, they may be signing blindly, not really understanding what is to be done to them.

17 When that happens, unpleasant surprises are likely. One man may find that the drug he is taking has debilitating side effects no one warned him of. He accuses his doctor of malpractice. A woman may

think she has agreed to exploratory surgery—and awake to find she has undergone a major amputation. She brings a malpractice suit. Some patients are simply disappointed that procedures do not turn out as well as expected. They too file for malpractice.

18 The growing number of malpractice suits alarmed HCPs. Some, convicted of delivering faulty care, had to pay hundreds of thousands of dollars in damages. Others tried to protect themselves by taking out malpractice insurance. That way, if they were found guilty of malpractice, the insurance companies would pay. But insurance executives were also alarmed by all the suits. They raised their rates so high that some doctors could not afford them. Doctors were beginning to find themselves in a predicament. How could they get out of it? Many decided the answer lay in the patients' rights movement.

A "Bill of Rights" for Patients

19 The patients' rights movement was new in the 1960s, but it caught on rapidly. Leaders of the movement included a number of HCPs as well as patients themselves. Their goal was to establish formal guidelines—a sort of "patients' bill of rights"—to define the relationship between patients and those who care for them. Such relationships, leaders of the movement believed, must be based on mutual respect. Not only must the patient respect the health professionals, but professionals must respect their patients. They must recognize patients as individual human beings with specific human needs and rights. It was that recognition that patients' rights activists sought to write into their guidelines.

20 An important milestone for the movement came in 1970. That year, the Joint Commission on the Accreditation of Hospitals put its weight behind the idea of a bill of rights for hospital patients. About 70 percent of this country's hospitals have commission accreditation. The group's 1970 action was to make adoption of a patients' bill of rights the first item on its list of standards. That meant no hospital could be accredited unless it had formally agreed to safeguard the rights of the people entrusted to its care.

21 Two years later, the American Hospital Association (AHA) took a similar action. Nearly all 7,000 American hospitals belong to the AHA. In 1972 the group drew up a twelve-point bill of rights and began urging its adoption.

22 The AHA document is very like the Joint Commission's. Both

emphasize two areas of rights. First is the right to give consent—informed consent—to treatment. Each declaration stresses the responsibility of HCPs and institutions to be sure patients thoroughly understand their conditions and the recommended course of treatment. The second area addressed in each document is the patient's right to privacy.

23 Adoption of the bills was a triumph for the patients' rights movement. But it was only a first step. The next, making the rights a reality, is less easily accomplished.

24 In the first place, the AHA did no more than make a recommendation. The Joint Commission went further, making adoption of patients' rights a standard that must be met for accreditation. However, no hospital is required to meet the Commission's standards. It can be to an institution's advantage to be accredited, since the federal government will then automatically repay it for some of the costs of treating its elderly patients. An unaccredited hospital may have to wait for a special inspection before payment is made. Still, a hospital that does not want to bother with a bill of rights can simply forego accreditation.

25 A second problem is that even if a hospital has an official bill of rights, no outside agency checks to see how well it is enforced. Even when HCPs are sincerely committed to enforcing patients' rights, they may find it difficult to do so. In some cases, they may even decide enforcement is unethical. That is because the ethics of informed consent and the right to privacy turn out to be much more complex than they appear to be on the surface.

Informed Consent—Is It Always Possible?

26 Consider consent to treatment and the sheer amount of information patients need if their consent is to be truly informed. If a disease is serious, its treatment is bound to be complicated. Even with the best intentions, doctors may not be able to make clear to patients exactly what they want to do.

27 At times the doctor may not fully understand a patient's condition and prospects. "Though anticoagulants, antibiotics, hypotensive agents, insulin, and steroids have been available for 15 to 40 years," writes one bioethicist, "many of their true effects are unknown. . . . [Doctors] are still uncertain about the best means of treatment for even such routine problems as a common cold, a sprained back, a fractured hip. . . ." How, then, can they be certain about the best way to treat a rare

cancer or to use a potent new drug? How can they fully inform a patient about something they themselves may not really understand?

28 Another problem can be that HCPs who believe in obtaining informed consent in general may question the wisdom of doing so in a particular case. Suppose a doctor is dealing with a person who is highly suggestible. Tell this patient that he might experience a certain symptom, and he promptly does. Say that his medicine may cause nausea, and he quickly gets sick. Such reactions may slow his recovery—or prevent it altogether. Is it going to help him to learn all the symptoms he *could* develop, all the complications surgery *can* lead to, all the *possible* side effects of the drugs prescribed for him? Would it really be unethical to gloss over the truth a little?

29 Other considerations can cloud the issue. A sick woman who is depressed to begin with may sink into profound despair if a doctor says her illness could be fatal. Her despair may keep her from cooperating fully with the doctor. Should the doctor phrase the diagnosis differently—make it sound less serious? Should the doctor lie a little for the patient's own good?

30 Another patient may be depressed, depressed enough to refuse treatment outright. He may prefer to die quickly, avoiding months of the therapy that even his own doctor warns will be painful and expensive. Then the doctor has observed the patient's rights—and cost him his life. Some patients who are confronted with catastrophic diagnoses may kill themselves. How moral is it to observe a patient's right to be informed if it leads directly to his death?

31 Questions like these have no simple answers. Some doctors think it is ethical to withhold information in some cases. Being less than totally honest is moral, they say, when it helps the patient in the long run. To support such a view, doctors may quote the oldest of medical maxims: "First, do no harm." Sometimes, they contend, the way to avoid doing harm is to ignore a patient's so-called right to the truth.

32 Other doctors reject this idea. To them, concealing the truth is no different from telling a direct lie. Both are unethical because both deny patients their human right to make free choices based on full knowledge of the facts. Honesty may hurt some patients at first, but dishonesty is a lasting moral wrong.

33 If the issue of informed consent turns out to be more involved than it appears at first, so does the matter of privacy. This right, the second covered in most hospital guidelines, encompasses two distinct

areas: physical privacy, and confidentiality—the right of patients to keep their personal and medical histories a secret between their doctors and themselves.

The Right to Privacy

34 In the past, physical privacy was not the issue it is today. Years ago, most patients were treated in the privacy of their own homes. Today, hospitalization is the rule. Many people in hospitals and other health-care facilities find physical privacy to be almost nonexistent.

35 In crowded institutions, it is virtually impossible for a patient to go alone into a room and close the door, confident of not being disturbed. In many places, several patients are jammed into a single room or forced to live together in large, barracks-like spaces. What goes on in one corner can be overheard and overseen from any other. At teaching hospitals, where medical students are trained, patients may have to undergo one examination after another as young doctors learn to "make the rounds."

36 Can anything be done about these invasions of privacy? Administrators at some institutions have placed limits on the number of examinations students may carry out on patients. Many HCPs have become more sensitive about providing a measure of privacy by placing screens around beds while examining or caring for patients. The patients' rights movement has been responsible for many such changes.

37 But the movement has not been able to do much to safeguard the other kind of privacy—the confidentiality of patient-doctor relations. Even as the movement was making it an issue, the last vestiges of confidentiality may have been vanishing from the American medical scene.

An End to Confidentiality

38 Patients must confide in their doctors. A doctor must know all a patient's symptoms, even the most intimate. He or she needs to know about other things too—drinking and smoking habits, diet and exercise patterns, whether the patient has irrational fears or deep depressions. Even the patient's sex life may be relevant to treatment.

39 Once, patients could trust doctors and other HCPs with such information. Most medical professionals deserved the trust. They valued their patients' privacy and felt a responsibility to respect their personal secrets.

40 Today, many do not. Tune into any TV soap opera, and you're

sure to see "doctors" and "nurses" chatting freely about this or that "patient." The gossip is essential on TV—it's one way soap opera writers keep their plots thickening. The problem is that many real-life HCPs seem little different from their TV counterparts. "In too many hospitals," says a spokesperson for one Boston hospital, "rounds may be conducted in hallways or elevators." Administrators at that hospital have posted signs: "Hospital staff are reminded that patient information should not be discussed in public areas."

41 The gossip isn't confined to hospitals. Dr. Lawrence K. Altman, medical correspondent for *The New York Times,* recounts a disturbing experience aboard an Amtrak train. In one car a young doctor was telling a companion about some of his cases. Absorbed by what he was saying, he failed to notice how loudly he was speaking. Soon, everyone in the vicinity knew all about his patients—including their names. "Some doctors," Dr. Altman concludes, "no longer value confidentiality as highly as virtually all doctors once did."

42 Why not? Again, part of the answer goes back to some of the ways medicine has changed. Once doctors knew most of their patients as individuals with secrets that were important to them. Now a busy doctor has so many patients that it's hard to get to know them personally. To present-day doctors, patients may not be so much distinct personalities as they are a collection of fascinating symptoms needing to be treated. Patients are medical challenges, not private human beings.

43 Nor is careless talk by doctors themselves the only threat to patient confidentiality. Our medical secrets are known to many—to doctors' receptionists, assistants, nurses, and billing clerks. The charts and records of hospital patients are open to many more: laboratory and x-ray technicians, dieticians, student doctors, residents and interns, three shifts of nurses and nurses' aides, volunteers, the men and women who program the computers in which hospital records are kept. According to the American College of Hospital Administrators, an average of seventy-five people in every major hospital have access to patient records. Any of these seventy-five people may spill a secret. If doctors cannot control their own gossiping, how can they expect to stop that of others?

44 But even if everyone with access to medical records were absolutely discreet, secrets would continue to leak out. Says Dr. Altman, "The insurance system alone has eliminated any remnant of confidentiality."

Other Threats to Confidentiality

45 Medical care in the United States is expensive, and only a multimillionaire could afford to pay for treatment of a long or serious illness. For the rest of us, paying our doctors' bills means relying on private medical insurance or such government funding programs as Medicare (for the elderly) or Medicaid (for the needy). No matter which patients turn to, their medical histories are opened up to another set of people.

46 Doctors are required to turn over their records to insurance companies upon request. This is reasonable. The companies have a right to know about the health of the people whose bills they pay. But how does opening the records affect the patient's right to privacy? If an insurance company finds out that someone has an alcohol problem, will his rates go up? His insurance be canceled? If he is insured through the place where he works, chances are his employer may learn about his alcoholism. Will that mean fewer promotions? The loss of his job?

47 When a person receives Medicaid or Medicare funds, the state or federal government gets into the act. Government inspectors, like insurance-company employees, may demand to view patient records. "What if Medicare finds out that you're a drug addict?" asks Robert Gelman, a lawyer for a congressional committee that has been studying the problem of disappearing patient confidentiality. "There's an incredible amount of information on everybody and we are losing control of it," Gelman warns.

48 Other factors threaten confidentiality. Government information about our health goes beyond Medicare and Medicaid records. In 1983, the federal Department of Health and Human Services adopted a "squeal rule." This rule applied to any federally funded clinic that gave out birth-control information. Under it, whenever such a clinic prescribed a birth-control method for a teen-aged girl, it was required to inform the girl's parents. It's easy to see how the squeal rule got its name!

49 Although officials at Health and Human Services defended their rule as an excellent way to improve teenagers' morals and help family members to communicate better, most HCPs condemned it as an unethical attack upon the right to privacy. After nearly a year of fierce public debate, the government abandoned its controversial regulation.

50 The squeal rule may have been an inappropriate attempt to legislate teenagers' morals, but other rules that require HCPs to report on certain of their patients may have a more compelling justification. "The right of privacy is not absolute," ruled a California court in one health-related case, "and in some cases is subordinate to the state's fundamental right to enact laws which promote public health, welfare and safety."

51 When might the public welfare be at stake? Suppose a doctor is called in to examine a patient with a gunshot wound. By law the doctor must report that patient to the police. That is because the injury may be evidence that a crime has taken place. If it has, the state has a compelling interest in bringing the criminal to justice.

52 Yet some HCPs question the ethics of obeying laws that require them to report such wounds. If they do obey, what happens to their patients' right to privacy? Suppose a doctor in a hospital emergency room finds that a patient is carrying heroin packaged for sale? Should he report that? On the one hand, selling heroin is a crime that endangers members of the public. On the other, the patient, by appearing in the emergency room, has entrusted himself to the doctor. Must the doctor betray that trust?

53 Or the doctor in question could be a psychiatrist. If one of the doctor's patients says she is going to kill her husband, what should the doctor do? Report it to the police? Warn the husband? Assume that the wife isn't really serious—that she is merely venting her aggression in a harmless way? What if a patient speaks of killing the president of the United States?

54 There is another—and new—threat to patient privacy today. That threat is the computer.

55 In June 1983, a computer systems manager at one New York City hospital discovered a problem in the hospital's computerized records. A computer that had been monitoring therapy for 250 patients had failed. In addition, about $1,500 worth of billing records had been destroyed. It was clear that some computer "hacker" had learned the system's password and broken into it. Unless the culprit were caught, all the hospital's patient records could be at risk. Private information might be stolen from those records, or false information added.

56 The culprit was caught—that time. But what if other hackers gain access to other medical records? American medicine is relying more and more upon computers. In two New York City hospital clin-

ics, for instance, computers are used to keep track of the pediatric care given to children in order to make that care more efficient. Dr. Richard K. Stone, chief of pediatrics at one of the hospitals, is enthusiastic about the program. "In one way," he says, "we are now looking over a doctor's shoulder with the help of a computer printout." Will the doctor be equally enthusiastic if unauthorized intruders also start peering over the doctor's shoulder? The threat that computer misuse poses to patient privacy is a bioethical issue that people in the medical community are just beginning to consider.

57 Where does all this—the disappearance of confidentiality, the problems of providing physical privacy, the dilemmas over informed consent—leave patients' rights? The good news is that the idea of patients' rights is now official policy at thousands of the institutions that care for the sick, the elderly, and the handicapped. The bad news is that, in the real world of practical, everyday considerations, many patients remain unable to truly enjoy those rights.

58 Both the good news and the bad are reflected in the most recent code of the American Medical Association. That code, adopted in 1981, instructs the physician to "respect the rights of patients." For the first time ever, the AMA has officially acknowledged that patients do have certain rights that need to be specifically spelled out. But the group's code continues to leave it up to each doctor to work out how—and how effectively—he or she will observe those rights.

Suggestions for Discussion and Writing

1. What changes in people's attitudes toward medical care have influenced the increase in malpractice suits, according to Weiss? Are there other causes that Weiss doesn't mention?

2. What is a patient's bill of rights? What does it usually include? What does it exclude? What rights does a patient give up if she or he signs this bill of rights?

3. There is a medical axiom, "First, do no harm." How does this apply to patients' rights? To doctors' responsibilities? Would you include placebos under the category of "doing no harm"?

4. How would you describe the organization of Weiss's essay? Why do you think she arranges her topics in this order?

Can you think of other arrangements of this material that might also be effective?

5. Some states, counties, and municipalities have enacted laws requiring hospitals to inform patients of their rights. Does your locality have such laws? What do they say? Assume you have been asked to argue in favor of such a law for your locality. What points would you bring up to community leaders?

Lester C. Thurow

The Ethical Costs of Health Care

LESTER C. THUROW was born in Livingston, Montana, in 1938, but has spent most of his professional life on the East Coast. A graduate of Oxford's Balliol College and of Harvard, he is a professor and former Dean of MIT's prestigious Sloan School of Management, and testifies frequently at government hearings on how economic policy affects various classes of our society. Among his recent books are *The Management Challenge: Japanese Views* (1985) and *The Zero-Sum Solution: Building a World-Class American Economy* (1985). This essay first appeared in *The New England Journal of Medicine* in 1985.

Although there is no magic formula for determining the precise limit on what a country can afford to spend on health care, there *is* a limit. Every dollar spent on health care is a dollar that cannot be spent on something else. No set of expenditures can rise faster than the gross national product forever. If the United States expects to increase its international competitiveness, the rate of growth of health-care expenditures must slow down to the rate of growth of the gross national product.

2 It is standard medical practice in the United States to continue

treatments until they yield no additional payoffs. But with the development of more and more elaborate techniques and devices that can slightly improve a diagnosis or briefly prolong life, the expenditures that end up being made before this traditional stopping point is reached have grown astronomically.

3 These new techniques require a shift in standard medical practice. Instead of stopping treatments when all benefits cease, physicians must stop treatments when marginal benefits are equal to costs. But where does this point lie? And who is to make the decision—the patient, the doctor, some third party? How do we as a society decide that we cannot afford a medical treatment that may marginally benefit someone?

4 Ethically, most Americans are simultaneously egalitarians and capitalists. This set of beliefs leads to an alarming chain reaction. A new and expensive treatment is developed. Since, as capitalists. Americans believe that individuals should be allowed to spend their money on whatever they wish, the wealthy are allowed to buy the treatment privately. People who cannot afford the treatment start to demand it. Being egalitarians, Americans do not have the political ability to say "no" to any person dying from a treatable disease, and so ways are found to pay for the treatment through private or public health insurance. As egalitarians, we feel we have to provide the treatment to everyone or deny it to everyone; as capitalists, we cannot deny it to those who can afford it. But since resources are limited, we cannot afford to give it to everyone.

5 As medical costs rise, it becomes less and less possible for us to live with our inconsistent ethical beliefs. At some time—and the time is now—the inconsistencies have to be sorted out.

6 Insurance has been the traditional solution for those who cannot afford to buy health care, but it is not the answer here. Insurance is an appropriate remedy in situations in which there is a small probability of a disaster that will incur large fixed losses. Fire insurance is the best example. Only a few of us will be unfortunate enough to have our houses burn down, and the maximal loss for any individual is fixed by the value of his house. We therefore pool our risks and compensate those who suffer losses. Companies make money by being good at estimating risks and choosy about whom they insure.

7 Although it is highly unlikely that any individual's house will burn down, it is almost certain that everyone will incur large health-

care expenditures. In this circumstance, insurance becomes not a pooling of small risks but an enormous distortion of incentives.

8 Each of us knows that our health-care expenditures in any given year have no impact on the following year's insurance rates. As a result, we have no incentive to restrain our expenditures. Insurance companies actually have an interest in increasing health-care spending, since they make money not by assessing risks and carefully selecting their policyholders but by taking a management fee that is usually a percentage of total expenditures. And doctors practicing in a fee-for-service system have an interest in prescribing services, since they raise their own income by doing so; moreover, because of insurance, doctors know they will not be directly raising costs for their patients if they recommend treatment. The result, not surprisingly, is exploding expenditures.

9 The health-care problem is not a federal or state budget problem. It is a social problem. The expenditures are the same regardless of whether the money is spent through the federal budget or private insurance. Somehow, we have to learn to say "no."

10 We can place greater reliance on market mechanisms, but if we do that we are saying that the capitalistic part of our ethics should dominate the egalitarian part. The market is often described as if it were a mechanism for limiting waste, but that is not its primary virtue. It is a mechanism for saying "no," but in a very inegalitarian way. Since the richest 20 percent of all U.S. households have 11 times as much income as the poorest 20 percent, any efficient market mechanism will give 11 times as much medical care to the top 20 percent as it gives to the bottom 20 percent.

11 The Reagan Administration's recent proposals for higher deductibles and prospective rather than retrospective payment clearly show the difficulties of solving the problem. To discourage the use of expensive health-care facilities, the government announced that Medicare will pay less and users must pay more. Private health-insurance companies responded by saying that they will sell insurance to cover what is not covered by the government—thus undercutting the whole purpose of the plan. Those who can afford coinsurance do not have to face the market incentives to use less health care. Those who cannot afford coinsurance must face those incentives. But are we really going to deny medical care to patients who cannot afford the necessary private payments?

12 Under a prospective payment system, hospitals are paid a fixed

amount for each case according to the disease diagnosed, not how much it actually cost to treat the patient. What is likely to result? Many hospitals will try to avoid admitting patients who are likely to be expensive to treat and who will not be able to pay for their treatment. Once again, this system leaves high-cost patients with limited means out in the cold.

13 These patients will be "dumped," either before or after admission, just as uninsured high-cost patients are now being dumped. No hospital wants to treat patients without money. Such patients generally end up in public, tax-supported hospitals—often municipal hospitals in big cities. But as city governments with their own budget problems attempt to restrain municipal hospital spending, treatment at such hospitals tends to become second-class. To deny that this will happen is to deny that markets are efficient.

14 Societies allow market mechanisms to work when buyers are knowledgeable or willing to live with their mistakes and when society is willing to distribute goods and services in accordance with the market's distribution of income. In the case of health care, neither of these necessary conditions exists.

15 We are not real believers in the free market if we are not willing to see some patients suffer the consequences when they cannot afford a treatment being provided to wealthier patients. If we cannot accept that, then, when push comes to shove, we simply will not let the market work.

16 Proponents of the market approach often forget that an egalitarian distribution of health care helps create social solidarity and a feeling of community. If health care is not part of the social glue that holds us together, what is?

17 If you are an egalitarian when it comes to medical care, and I confess that I am, what is the answer? One answer is to allow third-party payers to write the rules and regulations concerning what they will and will not pay for, and to prohibit their clients from buying services that are not permitted under the private or public insurance systems. This is essentially how the British have kept health-care spending at half the American level.

18 Such a procedure works, but it works clumsily, since no set of rules can be adjusted to the nuances of individual medical problems. It will be far better if American doctors develop procedures to help them decide when medicine is bad medicine—not simply when it has

absolutely no payoff or hurts the patient, but when the costs are not justified by the benefits.

19 The medical profession has norms concerning what constitutes bad medical practice. But those norms do not take into account cases in which high costs are not justified by minor expected benefits. If new standards are developed, and then defended in court against malpractice claims, a system of doctor-imposed cost controls might be put in place that would be much more flexible than any system imposed by third-party payers could be. If the medical profession fails to do this, sooner or later the United States will move to a system of third-party controls.

20 As a society, how much are we willing to spend (sacrifice) to prolong life? The easy answer is any amount, but that answer is neither true nor feasible. Like it or not, Americans are going to have to come to a consensus concerning the trade-off between the costs of medical services and the life-extending benefits that result.

Suggestions for Discussion and Writing

1. What questions does Thurow feel are most central to the debate over controlling medical costs? How does he define costs?

2. Explain this statement: "Ethically, most Americans are simultaneously egalitarians and capitalists. This set of beliefs leads to an alarming chain reaction."

3. What are the strengths and weaknesses of the market approach to controlling health-care costs?

4. Thurow uses few actual facts and figures in his argument. Why do you think he chose to use so few numbers? How does that choice affect you as readers of his argument?

5. Supposing you were appointed to a public panel charged with making decisions about what kinds of health care the government should pay for. What guidelines would you set up to shape your decisions? What rights and responsibilities would you want considered?

6 The Future of Science and Technology

"Brave new world!" When a Shakespearean heroine uttered this phrase, it was a summary of all the hopes people in Elizabethan England had for the discoveries of the future. By the time Aldous Huxley appropriated it in this century, it had become an ironic catch phrase for humanity's cynicism and distrust of what the future might hold. Somewhere between these extremes lies the truth—if we can get at it.

In this section we present five very different views of the future science and technology hold out to us. John McPhee looks at novel ways to provide energy and conserve natural resources. Freeman Dyson postulates about the future of space exploration, while O.B. Hardison wonders what the effects of a universal culture may be. Finally, Grant Fjermedal looks at the next "uncrossable" barrier we are preparing to cross: that between life and death. As you read these essays, ask yourself, "What do I think the future will hold? Are change and progress always good? In what directions do I want science and technology to take our culture? How will I participate in that future?"

John McPhee
Ice Pond

Born in New Jersey in 1931, JOHN MCPHEE graduated from Princeton in 1953 and worked for a time as a television script writer and at *Time* before becoming a professor of journalism at Princeton in 1964. Since that time he has established himself as one of America's leading non-fiction writers, specializing in studies of offbeat subjects and people who practice obscure or fading crafts very, very well. His books include *Levels of the Game* (1970), *Coming into the Country* (1977), *Basin and Range* (1981), *Rising from the Plains* (1986), and *Looking for a Ship* (1990).

McPhee's prose has been praised because "It enables McPhee to translate for the layman the mysteries that preoccupy professionals, be they athletes or engineers." McPhee himself says only, "What you hope is that some subject will interest you and then you will have to deal with it on its own terms. . . . Critics may think I should be doing things on a grander scale. . . . But fundamentally, I'm a working journalist and I've got to go out and work." This essay is from *Table of Contents* (1985).

Summer, 1981

At Princeton University, off and on since winter, I have observed the physicist Theodore B. Taylor standing like a mountaineer on the summit of what appears to be a five-hundred-ton Sno-Kone. Taylor now calls himself a "nuclear dropout." His has been, at any rate, a semicircular career, beginning at Los Alamos Scientific Laboratory, where, as an imaginative youth in his twenties, he not only miniaturized the atomic bomb but also designed the largest-yield fission bomb that had ever been exploded anywhere. In his thirties, he moved on to General Atomic, in La Jolla, to lead a project called Orion, his purpose being to construct a spaceship sixteen stories high and as voluminous as a college dormitory, in which he personally meant to take off from a Nevada basin and set a course for Pluto, with intermediate stops on Ganymede, Rhea, and Dione—ice-covered satellites of Jupiter and Saturn. The spaceship Orion, with its wide flat base, would resemble the nose of a bullet, the head of a rocket, the ogival hat of a bishop. It would travel at a hundred thousand miles an hour and be driven by two thousand fission bombs. Taylor's colleague Freeman Dyson meant to go along, too, extending spectacularly a leave of absence from the Institute for Advanced Study, in Princeton. The project was developing splendidly when the nuclear treaty of 1963 banned explosions in space and the atmosphere. Taylor quelled his dreams, and turned to a sombre subject. Long worried about the possibility of clandestine manufacture of nuclear bombs by individuals or small groups of terrorists, he spent his forties enhancing the protection of weapons-grade uranium and plutonium where it exists in private industries throughout the world. And now, in his fifties—and with the exception of his service as a member of the President's Commission on the Accident at Three Mile Island—he has gone flat-out full-time in pursuit of sources of energy that avoid the use of fission and of fossil fuel, one example of which is the globe of ice he has caused to be made in Princeton. "This isn't Ganymede," he informs me, scuffing big crystals under his feet. "But it's almost as exciting."

2 Taylor's hair is salt-and-peppery now but still stands in a thick youthful wave above his dark eyebrows and luminous brown eyes. He is tall, and he remains slim. What he has set out to do is to air-condition large buildings or whole suburban neighborhoods using less than ten per cent of the electricity required to cool them by conventional means, thereby saving more than ninety per cent of the oil that might be used to make the electricity. This way and that, he wants to take the "E" out of OPEC. The ice concept is simple. He grins and calls it "simple-minded—putting old and new ideas together in a technology appropriate to our time." You scoop out a depression in the ground, he explains—say, fifteen feet deep and sixty feet across—and line it with plastic. In winter, you fill it with a ball of ice. In summer, you suck ice water from the bottom and pump it indoors to an exchanger that looks something like an automobile radiator and cools air that is

flowing through ducts. The water, having picked up some heat from the building, is about forty-five degrees as it goes back outside, where it emerges through shower heads and rains on the porous ice. Percolating to the bottom, the water is cooled as it descends, back to thirty-two degrees. Taylor calls this an ice pond. A modest number of ice ponds could cool, for example, the District of Columbia, saving the energy equivalent of one and a half million barrels of oil each summer.

3 The initial problem was how to make the ice. Taylor first brooded about this some years ago when he was researching the theoretical possibilities of constructing greenhouses that would aggregately cover tens of millions of acres and solve the pollution problems of modern agriculture. The greenhouses had to be cooled. He thought of making ice in winter and using it in summer. For various regions, he calculated how much ice you would have to make in order to have something left on Labor Day. How much with insulation? How much without insulation? The volumes were small enough to be appealing. How to make the ice? If you were to create a pond of water and merely let it freeze, all you would get, of course, would be a veneer that would break up with the arrival of spring. Ice could be compiled by freezing layer upon layer, but in most places in the United States six or eight feet would be the maximum thickness attainable in an average winter, and that would not be enough. Eventually, he thought of artificial snow. Ski trails were covered with it not only in Vermont and New Hampshire but also in New Jersey and Pennsylvania, and even in North Carolina, Georgia, and Alabama. To make ice, Taylor imagined, one might increase the amount of water moving through a ski-resort snow machine. The product would be slush. In a pondlike receptacle, water would drain away from the slush. It could be pumped out and put back through the machine. What remained in the end would be a ball of ice.

4 Taylor had meanwhile become a part-time professor at Princeton, and on one of his frequent visits to the university from his home in Maryland he showed his paper ice ponds to colleagues at the university's Center for Energy and Environmental Studies. The Center spent a couple of years seeking funds from the federal government for an icepond experiment, but the government was not interested. In 1979, the Prudential Insurance Company of America asked the university to help design a pair of office buildings—to be built

just outside Princeton—that would be energy-efficient and innovative in as many ways as possible. Robert Socolow, a physicist who is the Center's director, brought Taylor into the Prudential project, and Taylor soon had funds for his snow machine, his submersible pumps, his hole in the ground.

5 At Los Alamos, when Taylor got together on paper the components of a novel bomb he turned over his numbers and his ideas to other people, who actually made the device. Had such a job been his to do, there would have been no bombs at all. His mind is replete with technology but innocent of technique. He cannot competently change a tire. He has difficulty opening doors. The university hired Don Kirkpatrick, a consulting solar engineer, to assemble and operate appropriate hardware, while unskilled laborers such as Taylor and Freeman Dyson would spread insulating materials over the ice or just stand by to comment.

6 "The first rule of technology is that no one can tell in advance whether a piece of technology is any good," Dyson said one day. "It will hang on things that are unforeseeable. In groping around, one wants to try things out that are quick and cheap and find out what doesn't work. The Department of Energy has many programs and projects—solar-energy towers and other grandiose schemes—with a common characteristic: no one can tell whether they're any good or not, and they're so big it will take at least five years and probably ten to find out. This ice pond is something you can do cheaply and quickly, and see whether it works."

7 A prototype pond was tried in the summer of 1980. It was dug beside a decrepit university storage building, leaky with respect to air and water, that had cinder-block walls and a flat roof. Size of an average house, there were twenty-four hundred square feet of space inside. Summer temperatures in the nineties are commonplace in New Jersey, and in musty rooms under that flat roof temperatures before the ice pond were sometimes close to a hundred and thirty. The 1980 pond was square—seventy-five feet across and fifteen feet deep. It contained a thousand tons of ice for a while, but more than half of that melted before insulation was applied: six inches of dry straw between sheets of polyethylene, weighed down with bald tires. Even so, the old building was filled most of the time from June to September with crisp October air. Something under seven tons of ice would melt away on a

hot day. Nonetheless, at the end of summer a hundred tons remained. "It's a nice alternative to fossil fuels," Robert Socolow commented. "It has worked too well to be forgotten."

8 The concept having been successfully tested, the next imperative was to refine the art—technically, economically, and aesthetically. "The point is to make it elegant this time," said Freeman Dyson, and, from its hexagonal concrete skirt to its pure-white reflective cover, "elegant" is the word for the 1981 pond. Concealing the ice is a tent-like Dacron-covered free-span steel structure with six ogival sides—a cryodesic dome—which seems to emerge from the earth like the nose of a bullet, the head of a rocket, the hat of a bishop. Lift a flap and step inside. Look up at the summit of a white tower under insulation. Five hundred tons of ice—fifty-eight feet across the middle—rise to a conical peak, under layers of polyethylene foam, sewn into fabric like enormous quilts. It is as if the tip of the Finsteraarhorn had been wrapped by Christo.

9 Taylor, up on the foam, completes his inspection of the ice within, whose crystals are jagged when they first fall from the snow machine, and later, like glacier ice, recrystallize more than once into spheres of increasing diameter until the ultimate substance is very hard and resembles a conglomerate of stream gravel. The U.S. Army's Cold Regions Research and Engineering Laboratory has cored it with instruments of the type used on glaciers in Alaska. Suspended from a girder high above Taylor's head and pointing at the summit of the ice is something that appears to be a small naval cannon with a big daisy stuck in its muzzle. This is SMI SnowStream 320, the machine that made the ice. In its days of winter operation, particles plumed away from it like clouds of falling smoke. Unlike many such machines, it does not require compressed air but depends solely on its daisy-petalled propeller blades of varying length for maximum effectiveness in disassembling water. "We are harvesting the cold of winter for use in the summer," Taylor says. "This is natural solar refrigeration, powered by the wind. Wind brings cold air to us, freezes the falling water, and takes the heat away. We are rolling with nature—trying to make use of nature instead of fighting it. That machine cost seven thousand dollars. It can make about eight thousand tons of ice in an average winter here in Princeton—for thirty-five dollars a hundred tons. A hundred tons is enough to air-condition almost any house, spring to fall. In the course of a winter, that machine could make ten

thousand tons of ice in Boston, seven thousand in Washington, D.C., fifteen thousand in Chicago, thirty thousand in Casper, Wyoming, fifty thousand in Minneapolis, and, if anybody cares, a hundred thousand tons of ice in Fairbanks. The lower the temperature, the more water you can move through the machine. We don't want dry snow, of course. Snow is too fluffy. We want slop. We want wet sherbet. At twenty degrees Fahrenheit, we can move fifty gallons a minute through the machine. The electricity that drives the snow machine amounts to a very small fraction of the electricity that is saved by the cooling capacity of the ice. In summer, electrical pumps circulate the ice water from the bottom of the pond for a few tenths of a cent a ton. The cost of moving air in ducts through the building is the same as in a conventional system and is negligible in comparison with the electrical cost of cooling air. We're substituting ice water made with winter winds for the cold fluid in a refrigerated-air conditioner, using less than a tenth as much electrical energy as a conventional air-conditioning system. Our goal is to make the over-all cost lower than the cost of a conventional system and use less than one-tenth of the energy. We're just about there."

10 The Prudential's new buildings—a hundred and thirty thousand square feet each, by Princeton's School of Architecture and Skidmore, Owings & Merrill—will be started this summer on a site a mile away. They are low, discretionary structures, provident in use of resources, durable, sensible, actuarial—with windows shaded just enough for summer but not too much for winter, with heat developing in a passive solar manner and brought in as well by heat pumps using water from the ground—and incorporating so many other features thrifty with energy that God will probably owe something to the insurance company after the account is totted up. An ice pond occupying less than half an acre can be expected to compound His debt.

11 A man who could devise atomic bombs and then plan to use them to drive himself to Pluto might be expected to expand his thinking if he were to create a little hill of ice. Taylor has lately been mulling the potentialities of abandoned rock quarries. You could fill an old rock quarry a quarter of a mile wide with several million tons of ice and then pile up more ice above ground as high as the Washington Monument. One of those could air-condition a hundred thousand homes. With all that volume, there would be no need for insulation.

You would build pipelines at least ten feet in diameter and aim them at sweltering cities, where heat waves and crime waves would flatten in the water-cooled air. You could make ice reservoirs comparable in size to New York's water reservoirs and pipe ice water to the city from a hundred miles away. After the water had served as a coolant, it would be fed into the city's water supply.

12 "You could store grain at fifty degrees in India," Taylor goes on. "We're exploring that. The idea is to build an aqueduct to carry an ice slurry from the foothills of the Himalayas down to the Gangetic plain. With an insulated cover over the aqueduct, the amount of ice lost in, say, two hundred miles would be trivial—if the aqueduct is more than ten feet across. In place of electric refrigeration, dairies could use ice ponds to cool milk. Most cheese factories could use at least fifty thousand tons of ice a year. If all the cheese factories in the United States were to do that, they alone would save, annually, about six million barrels of oil. When natural gas comes out of the earth, it often contains too much water vapor to be suitable for distribution. One way to get rid of most of the water is to cool the gas to forty degrees. If ice ponds were used to cool, say, half the natural gas that is produced in this country, they would save the equivalent of ten million barrels of oil each year. Massive construction projects, such as dams, use amazing amounts of electricity to cool concrete while it hardens, sometimes for as much as three years. Ice ponds could replace the electricity. Ice ponds could cool power plants more effectively than environmental water does, and therefore make the power plants more efficient. Ice would also get rid of the waste heat in a manner more acceptable than heating up a river. In places like North Dakota, you can make ice with one of these machines for a few cents a ton—and the coolant would be economically advantageous in all sorts of industrial processing."

13 Taylor shivers a little, standing on the ice, and, to warm himself, he lights a cigarette. "You could also use snow machines to freeze seawater," he continues. "As seawater freezes, impurities migrate away from it, and you are left with a concentrated brine rich in minerals, and with frozen water that is almost pure—containing so little salt you can't taste it. As seawater comes out of the snow machine and the spray is freezing in the air, the brine separates from the pure frozen water as it falls. To use conventional refrigeration—to use an electric motor to run a compressor to circulate Freon to freeze seawater—is

basically too costly. The cost of freezing seawater with a ski-slope machine is less than a hundredth the cost of freezing seawater by the conventional system. There are sixty-six pounds of table salt in a ton of seawater, almost three pounds of magnesium, a couple of pounds of sulphur, nearly a pound of calcium, lesser amounts of potassium, bromine, boron, and so forth. Suppose you had a ship making ice from seawater with snow machines that had been enlarged and adapted for the purpose. You would produce a brine about ten times as concentrated with useful compounds as the original seawater. It would be a multifarious ore. Subsequent extraction of table salt, magnesium, fertilizers, and other useful material from the brine would make all these products cheaper than they would be if they were extracted from unconcentrated seawater by other methods. The table salt alone would pay for the ship. You could separate it out for a dollar a ton. A ship as large as a supertanker could operate most of the year, shuttling back and forth from the Arctic to the Antarctic. At latitudes like that, you can make twenty times as much ice as you can in Princeton."

14 "What do you do with the ice?"

15 "Your options are to return it to the sea or to put it in a skirt and haul it as an iceberg to a place where they need fresh water. The Saudis and the French have been looking into harvesting icebergs in Antarctica and towing them to the Red Sea. Someone has described this as bringing the mountain to Muhammad. I would add that if you happen to live in a place like New York the mountain is right at your doorstep—all you have to do is make it. The cost of making fresh water for New York City with snow machines and seawater would be less than the cost of delivered water there now. Boston looks awfully good—twice as good as Princeton. Boston could make fresh water, become a major producer of table salt and magnesium and sulphur, and air-condition itself—in one operation. All they have to do is make ice. It would renew Boston. More than a hundred years ago, people cut ice out of ponds there and shipped it around Cape Horn to San Francisco. When this country was getting going, one of Boston's main exports was ice."

Suggestions for Discussion and Writing

1. What is an ice pond? Why could its development be important to our future?

2. Why does McPhee begin with the extended biography of Ted Taylor? What effect does this have on McPhee's readers?

3. Why does McPhee stress that engineers like Taylor and Dyson are "unskilled laborers" who "cannot competently change a tire"? What attitude toward his subjects does McPhee convey with details like these?

4. What factors might influence other companies and communities to develop alternate energy sources like ice ponds? Who do you think would support such projects? Who would oppose them?

5. What is the status of alternate energy source development in your community? What kinds of projects are under discussion, or under way? Why do you think big cities haven't endorsed Taylor's suggestions? Write a speech for a political candidate in your area that lays out her position on alternative energy.

Freeman Dyson

The Astrochicken

Born in 1923, FREEMAN DYSON is a scientist whose professional interests include mathematics, nuclear physics, rocket technology, and astrophysics. His personal passions include many of today's most pressing social and political issues, especially nuclear disarmament. He worked with J. Robert Oppenheimer, Edward Teller, and Theodore Taylor on many of our most significant nuclear energy research projects; this diversity has helped him propose compromises and clarifications for many issues where technology and public policy collide. Among his books are *Disturbing the Universe* (1979), *Values at War* (1983), *Weapons and Hope* (1984), *Infinite in All Directions* (1985), and *Origins of Life* (1985). This is a selection from *Infinite in All Directions*.

I am dreaming of the next space mission to explore the planet Uranus. I call the mission Uranus 2, and I imagine it arriving at Uranus in the year 2016. The Voyager fly-by was only a beginning. Like all good scientific missions, it raised more new questions than it answered old ones. A single fly-by is not enough. We need to explore Uranus comprehensively and thoroughly, to study its chemistry and meteorology and evolutionary history, to study the magnetic field and the rings, the topography and geology of each of the moons.

2 This is an ambitious program. It is unlikely that it can be accomplished by further missions in the style of Voyager. The Voyager fly-by comes close to the limit of what can be achieved by the technology of the 1970s. I am assuming that when we go back to Uranus next time, we will go with the technology of the twenty-first century. From Voyager to Uranus 2 is a big jump, as big a jump as we made from the primitive sounding rockets of the 1940s to Voyager. From the V-2 to Voyager was thirty years, and from Voyager to Uranus 2 will be another thirty years. It would have been difficult in 1946 to imagine Voyager, and it is difficult today to imagine Uranus 2. The difficulty in imagining the future comes from the fact that the important changes are not quantitative. The important changes are qualitative, not bigger and better rockets but new styles of architecture, new rules by which the game of exploration is played. It would have been difficult to imagine Voyager in 1946 because the concept of on-board reprogrammable software did not then exist. It is difficult to imagine Uranus 2 today because the concept of a biologically organized spacecraft does not yet exist.

3 The Uranus 2 mission is a good one to dream about. It is too far away to be reached by our existing technology, but not too far away to be achieved by young people alive today if they are not afraid to break new paths into the future. I am not saying that my version of the Uranus 2 mission will happen, or that it ought to happen. The real future of space science may turn out to be quite different. All I am saying is that my version of Uranus 2 is a possibility, and that it is not too soon to begin thinking about it. Uranus 2 will not be a one-shot mission like Voyager. If Uranus 2 flies as I imagine it, it will be only one among a large flock of similar birds flying out to various destinations all over the solar system. Before Uranus 2 flies, her cousins will already be exploring Mars and Jupiter and Saturn. Uranus 2 will be cheap enough to fly frequently. The spacecraft that flies the Uranus 2 mission will

be called Astrochicken. I want it to have a name like itself, cheerful and unpretentious.

4 The basic idea of Astrochicken is that the spacecraft will be small and quick. I do not believe that a fruitful future for space science lies along the path we are now following, with space missions growing larger and larger and fewer and fewer and slower and slower as the decades go by. I propose a radical step in the direction of smallness and quickness. Astrochicken will weigh a kilogram instead of Voyager's ton, and it will travel from Earth into orbit around Uranus in two years instead of Voyager's nine. The spacecraft must be far more versatile than Voyager. It must land on each of Uranus' moons, roam around on their surfaces, see where it is going, taste the stuff it is walking on, take off into space again, and navigate around Uranus until it decides to make a landing somewhere else. To do all this with a l-kilogram spacecraft sounds crazy to people who have to work and plan within the constraints of today's technology. Perhaps it will still be crazy in 2016. Perhaps not. I am dreaming of the new technologies, which might make such a crazy mission possible.

5 Three kinds of new technology are needed. All three are likely to become available for use by the year 2016. All three are already here in embryonic form and are advanced far enough to have names. Their names are genetic engineering, artificial intelligence and solar-electric propulsion. Genetic engineering is fundamental. It is the essential tool required in order to design a l-kilogram spacecraft with the capabilities of Voyager. Astrochicken will not be built, it will be grown. It will be organized biologically and its blueprints will be written in the convenient digital language of DNA. It will be a symbiosis of plant and animal and electronic components. The plant component has to provide a basic life-support system using closed-cycle biochemistry with sunlight as the energy source. The animal component has to provide sensors and nerves and muscles with which it can observe and orient itself and navigate to its destination. The electronic component has to receive instructions from Earth and transmit back the results of its observations. During the next thirty years we will be gaining experience in the art of designing biological systems of this sort. We will be learning how to coordinate the three components so that they work smoothly together.

6 Artificial intelligence is the tool required to integrate the animal and electronic components into a working symbiosis. If the inte-

gration is successful, Astrochicken could be as agile as a hummingbird with a brain weighing no more than a gram. The information-handling apparatus is partly neural and partly electronic. An artificial intelligence machine is a computer designed to function like a brain. A computer of this sort will be made compatible with a living nervous system, so that information will flow freely in both directions across the interface between neural and electronic circuits.

7 The third new technology required for Uranus 2 is solar-electric propulsion. To get from Earth to Uranus in two years requires a speed of 50 kilometers per second, too fast for any reasonable multistage chemical rocket. It is also too fast for solar sails. Nuclear propulsion of any kind is impossible in a 1-kilogram spacecraft. Solar-electric propulsion is the unique system which can economically give a high velocity to a small payload. In this system, solar energy is collected by a large, thin antenna and converted with modest efficiency into thrust. The spacecraft carries a small ion-jet motor which uses propellant sparingly and gives an acceleration of the order of a milligee.

8 Nobody has yet done the careful engineering development to demonstrate that the energy of sunlight can be converted into thrust with a power-to-weight ratio of 1 kilowatt per kilogram. That is what Uranus 2 needs. But solar-electric propulsion is probably an easier technology to develop than genetic engineering and artificial intelligence. Since I am talking science fiction, I shall assume that all three technologies will be available for our use in 2016. I can then give a rough sketch of the Uranus 2 mission.

9 The mission begins with a conventional launch taking the spacecraft from Earth into orbit. Since the spacecraft weighs only 1 kilogram, it can easily ride on any convenient launcher. During the launch, the spacecraft is packaged into a compact shape, and the biological components are busy reorganizing themselves for life in space. During this phase the spacecraft is a fertilized egg, externally inert but internally alive, waiting for the right moment to emerge in the shape of an Astrochicken. After it is in a low Earth orbit, it will emerge from its package and deploy the life-support apparatus needed for survival in space. It will deploy, or grow, a thin-film solar collector. The collector weighs 100 grams and collects sunlight from an area of 100 square meters. It feeds a kilowatt of power into the little ion-drive engine which sends the spacecraft on its way with a milligee

acceleration sustained for several months. This is enough to escape from Earth's gravity and arrive at Uranus within two years. The same 100-square-meter collector serves as a radio antenna for two-way communication with Earth. This is ten times the area of the Voyager high-gain antenna. For the same rate of information transmitted, the transmitter power of Astrochicken can be ten times smaller than Voyager, 2 watts instead of 20 watts.

10 The spacecraft arrives at Uranus at 50 kilometers per second and grazes the outer fringe of the Uranus atmosphere. The 100-square-meter solar collector now acts as an efficient atmospheric brake. Because the collector is so light, it is not heated to extreme temperatures as it decelerates. The peak temperature turns out to be about 800 Celsius or 1500 Fahrenheit. The atmospheric braking lasts for about half a minute and produces a peak deceleration of 100 gees. The spacecraft leaves Uranus with speed reduced to 20 kilometers per second and passes near enough to one of the moons to avoid hitting Uranus again. It is then free to navigate around at leisure among the moons and rings. The solar-electric propulsion system, using the feeble sunlight at Uranus, is still able to give the spacecraft an acceleration of a tenth of a milligee, enough to explore the whole Uranus system over a period of a few years.

11 The spacecraft must now make use of its biological functions to refuel itself. First it navigates to one of the rings and browses there, eating ice and hydrocarbons and replenishing its supply of propellant. If one ring tastes bad it can try another, moving around until it finds a supply of nutrients with the right chemistry for its needs. After eating its fill, it will use its internal metabolic processes with the input of energy from sunlight to convert the food into chemical fuels. Chemical fuels are needed for jumping onto moons and off again. Solar-electric propulsion gives too small a thrust for jumping. The spacecraft carries a small auxiliary chemical rocket system for this purpose. We know that a chemical rocket system is biologically possible, because there exists on the Earth a creature called the Bombardier beetle which uses a chemical rocket to bombard its enemies with a scalding jet of hot liquid. It manufactures chemical fuels within its body and combines them in its rocket chamber to produce the scalding jet. Astrochicken will borrow its chemical rocket system from the Bombardier beetle. The Bombardier beetle system will give it the ability to accelerate with short bursts of high thrust to escape

from the feeble gravity of the Uranus moons. The spacecraft may also prefer to use the Bombardier beetle system for jumping quickly from one place to another on a moon rather than walking laboriously over the surface. While living on the surface of a moon, the Astrochicken will continue to eat and to keep the Bombardier beetle fuel tanks filled. From time to time it will transmit messages to Earth informing us about its adventures and discoveries.

12 That is not the end of my dream, but it is the end of my chapter. I have told enough about the Uranus 2 mission to give the flavor of it. The underlying idea of Uranus 2 is that we should apply to the development of technology the lessons which nature teaches us in the history of the evolution of life. Birds and dinosaurs were cousins, but birds were small and agile while dinosaurs were big and clumsy. Big mainframe computers, nuclear power stations and Space Shuttles are dinosaurs. Microcomputers, STIG gas turbines, Voyager and Astrochicken are birds. The future belongs to the birds. The JPL engineers now have their dreams on board the Voyager speeding on its way to Neptune. I hope the next generation of engineers will have their dreams riding on Uranus 2 in 2016.

Suggestions for Discussion and Writing

1. What kind of creature will the Astrochicken be? Will it be fair to call it a living creature?

2. What benefits will humanity gain by sending the Astrochicken to Uranus? What might the costs be? Would Walters or Thurow point out other costs that Dyson doesn't mention? Why do you think Dyson left such considerations out of this essay?

3. Explain what Dyson means by this statement: "The future belongs to the birds."

4. Do you think Dyson is serious in suggesting a chicken as the space vehicle of the future? What kind of approach to science is he taking? Is this a genuine scientific proposal? What would Thomas or Judson say about Dyson's tone and attitude?

5. What directions do you think humanity's exploration of space will take? What factors might shape this exploration? Do you think someday humanity will "boldly go where no one has gone before?" If you are a science fiction fan, you might want to speculate about the points at which science fiction might intersect with reality.

O. B. Hardison

Disappearing Through the Skylight

O. B. HARDISON, born in 1928 in San Diego, was trained as a scholar of Renaissance literature and taught at the universities of Tennessee, North Carolina-Chapel Hill, Princeton, and Georgetown. But he is most remembered as the dynamic director of Washington, D.C.,'s Folger Shakespeare Library, from 1969 to 1983. There he was not only a successful fundraiser and administrator, but also found novel and diverse ways to bring Shakespeare to more people and generally to make "high-brow" literature more accessible to the American public.

In his later years, Hardison turned to writing about the problems of human identity in the modern world; his later works include *Toward Freedom and Dignity: the Humanities and the Idea of Humanity* (1972) and *Entering the Maze: Identity and Change in Modern Culture* (1981). This is taken from his last book, *Disappearing through the Skylight: Culture and Technology in the Twentieth Century* (1989); Hardison died in 1990.

Science is committed to the universal. A sign of this is that the more successful a science becomes, the broader the agreement about its basic concepts: there is not a separate Chinese or American or Soviet thermodynamics, for example; there is simply thermodynamics. For several decades of the twentieth century there was a Western and a Soviet genetics, the latter associated with Lysenko's theory that environmental stress *can* produce genetic mutations. Today Lysenko's theory is discredited, and there is now only one genetics.

2 As the corollary of science, technology also exhibits the universalizing tendency. This is why the spread of technology makes the world look ever more homogeneous. Architectural styles, dress styles, musical styles—even eating styles—tend increasingly to be world styles. The world looks more homogeneous because it *is* more homogeneous. Children who grow up in this world therefore experience it as a sameness rather than a diversity, and because their identities are shaped by this sameness, their sense of differences among cultures and individuals diminishes. As buildings become more alike, the people who inhabit the buildings become more alike. The result is described precisely in a phrase that is already familiar: the disappearance of history.

3 The automobile illustrates the point with great clarity. A technological innovation like streamlining or all-welded body construction may be rejected initially, but if it is important to the efficiency or economics of automobiles, it will reappear in different ways until it is not only accepted but universally regarded as an asset. Today's automobile is no longer unique to a given company or even to a given national culture. Its basic features are found, with variations, in automobiles in general, no matter who makes them.

4 A few years ago the Ford Motor Company came up with the Fiesta, which it called the "World Car." Advertisements showed it surrounded by the flags of all nations. Ford explained that the cylinder block was made in England, the carburetor in Ireland, the transmission in France, the wheels in Belgium, and so forth.

5 The Fiesta appears to have sunk without a trace. But the idea of a world car was inevitable. It was the automotive equivalent of the International Style. Ten years after the Fiesta, all of the large automakers were international. Americans had plants in Europe, Asia, and South America, and Europeans and Japanese had plants in America and South America, and in the Soviet Union (Fiat workers refreshed themselves with Pepsi-Cola). In the fullness of time international automakers will have plants in Egypt and India and the People's Republic of China.

6 As in architecture, so in automaking. In a given cost range, the same technology tends to produce the same solutions. The visual evidence for this is as obvious for cars as for buildings. Today, if you choose models in the same price range, you will be hard put at 500 paces to tell one make from another. In other words, the specifically

American traits that lingered in American automobiles in the 1960s—traits that linked American cars to American history—are disappearing. Even the Volkswagen Beetle has disappeared and has taken with it the visible evidence of the history of streamlining that extends from D'Arcy Thompson to Carl Breer to Ferdinand Porsche.

7 If man creates machines, machines in turn shape their creators. As the automobile is universalized, it universalizes those who use it. Like the World Car he drives, modern man is becoming universal. No longer quite an individual, no longer quite the product of a unique geography and culture, he moves from one climate-controlled shopping mall to another, from one airport to the next, from one Holiday Inn to its successor three hundred miles down the road; but somehow his location never changes. He is cosmopolitan. The price he pays is that he no longer has a home in the traditional sense of the word. The benefit is that he begins to suspect home in the traditional sense is another name for limitations, and that home in the modern sense is everywhere and always surrounded by neighbors.

8 The universalizing imperative of technology is irresistible. Barring the catastrophe of nuclear war, it will continue to shape both modern culture and the consciousness of those who inhabit that culture.

9 This brings us to art and history again. Reminiscing on the early work of Francis Picabia and Marcel Duchamp, Madame Gabrielle Buffet-Picabia wrote of the discovery of the machine aesthetic in 1949: "I remember a time. . . when every artist thought he owed it to himself to turn his back on the Eiffel Tower, as a protest against the architectural blasphemy with which it filled the sky. . . . The discovery and rehabilitation of. . . machines soon generated propositions which evaded all tradition, above all, a mobile, *extra* human plasticity which was absolutely new. . . ."

10 Art is, in one definition, simply an effort to name the real world. Are machines "the real world" or only its surface? Is the real world that easy to find? Science has shown the insubstantiality of the world. It has thus undermined an article of faith: the thingliness of things. At the same time, it has produced images of orders of reality underlying the thingliness of things. Are images of cells or of molecules or of galaxies more or less real than images of machines? Science has also produced images that are pure artifacts. Are images of self-squared dragons more or less real than images of molecules?

11 The skepticism of modern science about the thingliness of things implies a new appreciation of the humanity of art entirely consistent with Kandinsky's observation in *On the Spiritual in Art* that beautiful art "springs from inner need, which springs from the soul." Modern art opens on a world whose reality is not "out there" in nature defined as things seen from a middle distance but "in here" in the soul or the mind. It is a world radically emptied of history because it is a form of perception rather than a content.

12 The disappearance of history is thus a liberation—what Madame Buffet-Picabia refers to as the discovery of "a mobile extra-human plasticity which [is] absolutely new." Like science, modern art often expresses this feeling of liberation through play—in painting in the playfulness of Picasso and Joan Miró and in poetry in the non-sense of Dada and the mock heroics of a poem like Wallace Stevens's "The Comedian as the Letter C."

13 The playfulness of the modern aesthetic is, finally, its most striking—and also its most serious and, by corollary, its most disturbing—feature. The playfulness imitates the playfulness of science that produces game theory and virtual particles and black holes and that, by introducing human growth genes into cows, forces students of ethics to reexamine the definition of cannibalism. The importance of play in the modern aesthetic should not come as a surprise. It is announced in every city in the developed world by the fantastic and playful buildings of postmodernism and neomodernism and by the fantastic juxtapositions of architectural styles that typify collage city and urban adhocism.

14 Today modern culture includes the geometries of the International Style, the fantasies of façadism, and the gamesmanship of theme parks and museum villages. It pretends at times to be static but it is really dynamic. Its buildings move and sway and reflect dreamy visions of everything that is going on around them. It surrounds its citizens with the linear sculpture of pipelines and interstate highways and high-tension lines and the delicate virtuosities of the surfaces of the Chrysler Airflow and the Boeing 747 and the lacy weavings of circuits etched on silicon, as well as with the brutal assertiveness of oil tankers and bulldozers and the Tinkertoy complications of trusses and geodesic domes and lunar landers. It abounds in images and sounds and values utterly different from those of the world of natural things seen from a middle distance.

15 It is a human world, but one that is human in ways no one expected. The image it reveals is not the worn and battered face that stares from Leonardo's self-portrait, much less the one that stares, bleary and uninspired, every morning from the bathroom mirror. These are the faces of history. It is, rather, the image of an eternally playful and eternally youthful power that makes order whether order is there or not and that having made one order is quite capable of putting it aside and creating an entirely different one the way a child might build one structure from a set of blocks and then without malice and purely in the spirit of play demolish it and begin again. It is an image of the power that made humanity possible in the first place.

16 The banks of the nineteenth century tended to be neoclassic structures of marble or granite faced with ponderous rows of columns. They made a statement: "We are solid. We are permanent. We are as reliable as history. Your money is safe in our vaults."

17 Today's banks are airy structures of steel and glass, or they are storefronts with slot-machinelike terminals, or trailers parked on the lots of suburban shopping malls.

18 The vaults have been replaced by magnetic tapes. In a computer, money is sequences of digital signals endlessly recorded, erased, processed, and reprocessed, and endlessly modified by other computers. The statement of modern banks is "We are abstract like art and almost invisible like the Crystal Palace. If we exist at all, we exist as an airy medium in which your transactions are completed and your wealth increased."

19 That, perhaps, establishes the logical limit of the modern aesthetic. If so, the limit is a long way ahead, but it can be made out, just barely, through the haze over the road. As surely as nature is being swallowed up by the mind, the banks, you might say, are disappearing through their own skylights.

Suggestions for Discussion and Writing

1. One of Hardison's sources says that creative work "springs from inner need, which springs from the soul." Is this true only for the arts, or do you think it could apply to the sciences as well? In what ways? What would Oppenheimer have to say about Hardison's argument?

2. Is the universalizing power of technology a good thing, according to Hardison? In what ways? What might be the hidden costs of this movement away from diversity?

3. Much of Hardison's essay proceeds by metaphor. What are the key metaphors in this essay? How do they affect the readers? Are there other metaphors you can think of that Hardison might have used?

4. What do you think Hardison means by "the thingliness of things"? Why do you think he chose to use such an unusual word? What effects do unusual language choices like this have on readers?

5. Assume you are to appear on a cable TV debate show to discuss Hardison's statement that "The price (humanity) pays (for becoming universal) is that (it) no longer has a home in the traditional sense of the word. The benefit is that (humanity) begins to suspect home in the traditional sense is another name for limitations, and that home in the modern sense is everywhere and always surrounded by neighbors." What position would you take in the debate? What kinds of examples and metaphors would you use to make your case?

Grant Fjermedal
Artificial Intelligence

GRANT FJERMEDAL is the author of a book on monoclonal antibodies and cancer treatments, *Magic Bullets* (1984); *The Tomorrow Makers: A Brave New World of Living-Brain Machines* (1986); and *New Horizons in Amateur Astronomy* (1988). This is an excerpt from *The Tomorrow Makers.*

I'm sure that Hans Moravec is at least as sane as I am, but he certainly brought to mind the classic mad scientist as we sat in his fifth-floor office at Carnegie-Mellon University on a dark and stormy night. It was nearly midnight, and he mixed for each of us a bowl of chocolate milk and Cheerios, with slices of banana piled on top.

2 Then, with banana-slicing knife in hand, Moravec, the senior research scientist at Carnegie-Mellon's Mobile Robot Laboratory, outlined for me how he could create a robotic immortality for Everyman, a deathless universe in which life would go on forever. By creating computer copies of our minds and transferring, or downloading, this program into robotic bodies, Moravec explained, humans could survive for centuries.

3 "You are in an operating room. A robot brain surgeon is in attendance. . . . Your skull but not your brain is anesthetized. You are fully conscious. The surgeon opens your braincase and peers inside." This is how Moravec described the process in a paper he wrote called "Robots That Rove." The robotic surgeon's "attention is directed at a small clump of about one hundred neurons somewhere near the surface. Using high-resolution 3-D nuclear-magnetic-resonance holography, phased-array radio encephalography, and ultrasonic radar, the surgeon determines the three-dimensional structure and chemical makeup of that neural clump. It writes a program that models the behavior of the clump and starts it running on a small portion of the computer sitting next to you."

4 That computer sitting next to you in the operating room would in effect be your new brain. As each area of your brain was analyzed and simulated, the accuracy of the simulation would be tested as you pressed a button to shift between the area of the brain just copied and the simulation. When you couldn't tell the difference between the original and the copy, the surgeon would transfer the simulation of your brain into the new, computerized one and repeat the process on the next area of your biological brain.

5 "Though you have not lost consciousness or even your train of thought, your mind—some would say soul—has been removed from the brain and transferred to a machine," Moravec said, "In a final step your old body is disconnected. The computer is installed in a shiny new one, in the style, color, and material of your choice."

6 As we sat around Moravec's office I asked what would become of the original human body after the downloading. "You just don't bother waking it up again if the copying went successfully," he said. "It's so messy. Humans have got so many problems that you might just want to leave it retired. You don't take your junker car out if you've

got a new one."

7 Moravec's idea is the ultimate in life insurance: Once one copy of the brain's contents has been made, it will be easy to make multiple backup copies, and these could be stashed in hiding places around the world, allowing you to embark on any sort of adventure without having to worry about aging or death. As decades pass into centuries you could travel the globe and then the solar system and beyond—always keeping an eye out for the latest in robotic bodies into which you could transfer your computer mind.

8 If living forever weren't enough, you could live forever several times over by activating some of your backup copies and sending different versions of yourself out to see the world. "You could have parallel experiences and merge the memories later," Moravec explained.

9 In the weeks and months that followed my stay at Carnegie-Mellon, I was intrigued by how many researchers seemed to believe downloading would come to pass. The only point of disagreement was *when*—certainly a big consideration to those of us still knocking around in mortal bodies. Although some of the researchers I spoke with at Carnegie-Mellon, MIT, and Stanford and in Japan thought that downloading was still generations away, there were others who believed achieving robotic immortality was imminent and seemed driven by private passions never to die.

10 The significance of the door Moravec is trying to open is not lost on others. Olin Shivers, a Carnegie-Mellon graduate student who works closely with Moravec as well as with Allen Newell, one of the founding fathers of artificial intelligence, told me, "Moravec wants to design a creature, and my professor Newell wants to design a creature. We are all, in a sense, trying to play God."

11 At MIT I was surprised to find Moravec's concept of downloading given consideration by Marvin Minsky, Donner Professor of Science and another father of artificial intelligence. Minsky is trying to learn how the billions of brain cells work together to allow a person to think and remember. If he succeeds, it will be a big step toward figuring out how to join perhaps billions of computer circuits together to allow a computer to receive the entire contents of the human mind.

12 "If a person is like a machine, once you get a wiring diagram of how he works, you can make copies," Minsky told me.

13 Although Minsky doesn't think he'll live long enough to down-

load (he's fifty-seven now), he would consider it. "I think it would be a great thing to do," he said. "I've spent a long time learning things, and I'd hate to see it all go away."

14 Minsky also said he would have no qualms about waving goodbye to his human body and taking up residence within a robot. "Why not avoid getting sick and things like that?" he asked. "It's hard to see anything against it. I think people will get fed up with bodies after a while. Then you'll have another population problem: You'll have all the people of the past, as well as the new ones."

15 Another believer is Danny Hillis, one of Minsky's Ph.D. students and the founding scientist of Thinking Machines, a Cambridge-based company that is trying to create the kind of computer that might someday receive the contents of a brain. During my research several computer scientists would point to Hillis's connection machine as an example of a new order of computer architecture, one that's comparable to the human brain. (Hillis's connection machine doesn't have one large central processing unit as other computers do but a network of 64,000 small units—roughly analogous in concept, if not in size, to the brain's network of 40 billion neuronal processing units.)

16 "I've added up the things I want to do in my life, and it's about fifteen hundred years' worth of stuff," Hillis, now twenty-eight, told me one day as we stood out on the sixth-floor sun deck of the Thinking Machines building. "I enjoy having a body as much as anyone else does, but if it's a choice between downloading into a computer—even one that's stuck in a room someplace—and still being able to think versus just dying, I would certainly take that opportunity to think."

17 Gerald J. Sussman, a thirty-six-year-old MIT professor and a computer hacker of historic proportions, expressed similar sentiments. "Everyone would like to be immortal. I don't think the time is quite right, but it's close. I'm afraid, unfortunately, that I'm in the last generation to die."

18 "Do you really think that we're that close?" I asked.

19 "Yes," he answered, which reminded me of something Moravec had written not too long ago: "We are on a threshold of a change in the universe comparable to the transition from nonlife to life."

Suggestions for Discussion and Writing

1. What does Fjermedal mean by "robotic immortality"? How would such immortality be achieved? How feasible is the technology for achieving this immortality?

2. In the third paragraph, Fjermedal uses a great deal of technical language called "jargon" to describe how brains might be replicated. Why do you think he uses this language? What is its effect on readers?

3. Is duplication of the brain the same thing as recreating life? Is "being able to think" really the opposite of "dying"? Is this a technological or ethical question?

4. What problems does Fjermedal foresee with "downloading"? Can you think of any he does not mention?

5. The essay ends with a startling quotation from Professor Moravec. What does that quotation mean? Why do you think Fjermedal chose to end the essay with this quote? What would Brownowski or Bethell have to say about this prediction?

Acknowledgements

BETHEL: "Agnostic Evolutionists" by Tom Bethel. Copyright © 1985 by Harper's Magazine. All rights reserved. Reprinted from the February issue by special permission.

BOK: "A Critical View of Astrology" by Bart J. Bok. First appeared in *The Humanist* issue of September/October 1975 and is reprinted by permission.

BOK: "Placebos" from *Lying: Moral Choice in Public and Private Life* by Sissela Bok. Copyright © 1978 by Sissela Bok. Reprinted by permission of Pantheon Books, a division of Random House, Inc.

BROWNOWSKI: "The Nature of Scientific Reasoning" by Jacob Brownowski. Copyright © 1956, 1965, by Jacob Brownowski, and renewed 1984 by Rita Brownowski. Reprinted by permission of Julian Messner, a division of Simon & Schuster, Inc.

DYSON: "The Astrochicken" from *Infinite in All Directions* by Freeman Dyson. Copyright © 1988 by Freeman Dyson. Reprinted by permission of HarperCollins Publishers.

EINSTEIN: "$E = mc^2$" by Albert Einstein. From *Science Illustrated* Magazine. Copyright © 1946.

EISELEY: "Science and the Sense of the Holy" from *The Star Thrower* by Loren Eiseley. Copyright © 1978 by the estate of Loren C. Eiseley, Mabel L. Eiseley, Executrix. Reprinted by permission of Times Books, a division of Random House, Inc.

ELDREDGE: "Creationism Isn't Science" by Niles Eldredge. From *The New Republic*, April 6, 1981. Reprinted by permission of *The New Republic*, copyright 1981 by The New Republic, Inc.

FJERMEDAL: "Artificial Intelligence" by Grant Fjermedal. Reprinted with the permission of Macmillan Publishing Company from *The Tomorrow Makers: A Brave New World of Living-Brain Machines* by Grant Fjermedal. Copyright © 1986 by Grant Fjermedal.

GOULD: "Our Alloted Lifetimes" by Stephen Jay Gould. With permission from *Natural History*, Vol. 86, No. 7; Copyright the American Museum of Natural History, 1977.

HARDISON: "Disappearing Through the Skylight" by O.B. Hardison, Jr. Copyright © 1989 by O.B. Hardison, Jr. Used by permission of Viking Penguin, a division of Penguin Books USA Inc.

JASPERS: "Is Science Evil?" by Karl Jaspers. Reprinted from

Commentary, March 1950, by permission; all rights reserved.

JUDSON: "The Rage to Know" by Horace Freeland Judson. Reprinted from *The Atlantic Monthly*, August 1980. Copyright 1985 by Horace Freeland Judson.

LOPEZ: "Encounter on the Tundra" by Barry Lopez. Reprinted with the permission of Charles Scribner's Sons, an imprint of Macmillan Publishing Company from *Arctic Dreams* by Barry Lopez. Copyright © 1986 by Barry Holstrum Lopez.

MARANTO: "Genetic Engineering: Hype, Hubris, and Haste" by Gina Maranto. Copyright © 1986 by *Discover*. Reprinted by permission.

MARSHALL: "When Commerce and Academe Collide" by Eliot Marshall from *Science*, April 13, 1990. Copyright © 1990 by the AAAS.

McPHEE: "Ice Pond" from *Table of Contents* by John McPhee. Copyright © 1985 by John McPhee. Reprinted by permission of Farrar, Straus & Giroux, Inc.

MEE: "The Summer Before Salk" by Charles L. Mee. Copyright © 1983 by Charles L. Mee, Jr. Reprinted by permission of the Wallace Literary Agency, Inc. First appeared in *Esquire*, December, 1983.

MEYER: "Do Lie Detectors Lie?" by Alfred Meyer, from *Science* Magazine, June 1982. Copyright © 1982 by Alfred Meyer. Reprinted by permission of the author.

MILGRAM: "The Perils of Obedience" by Stanley Milgram. First appeared in *Harper's*, Vol. 247, No. 1483, December 1973, pp. 62-66, 75-77. Copyright Alexandra Milgram. Reprinted by permission of Alexandra Milgram.

PETRUNKEVITCH: "The Spider and the Wasp," by Alexander Petrunkevitch. Reprinted with permission. Copyright © 1952 by Scientific American, Inc. All rights reserved.

THOMAS: "Debating the Unknowable," by Dr. Lewis Thomas. First Published in *The Atlantic Monthly* July 1981. Reprinted by permission of the author.

THUROW: "The Ethical Costs of Health Care" from "Learning to Say No" by Lester C. Thurow. Copyright © 1985 by *The New England Journal of Medicine*. Reprinted by permission.

WALTERS: "Ethical Issues in the Prevention and Treatment of HIV Infection and AIDS," by LeRoy Walters. Published in *Science* 2/5/88, volume 239, beginning on page 597. Copyright © 1988 by the AAAS. Reprinted by permission of the AAAS and LeRoy Walters.

WEISS: "The Rights of Patients" chapter 2 of *Bioethics* by Ann E. Weiss. Copyright © 1985 by Enslow Publishers. Reprinted by permission of Enslow Publishers, Inc.